AF790332

EAA Series

EAA series is successor of the EAA Lecture Notes and supported by the European Actuarial Academy (EAA GmbH), founded on the 29 August, 2005 in Cologne (Germany) by the Actuarial Associations of Austria, Germany, the Netherlands and Switzerland. EAA offers actuarial education including examination, permanent education for certified actuaries and consulting on actuarial education.

actuarial-academy.com

For further titles published in this series, please go to
http://www.springer.com/series/7879

Andreas E. Kyprianou

Gerber–Shiu Risk Theory

Springer

Andreas E. Kyprianou
Department of Mathematical Sciences
University of Bath
Bath, United Kingdom

ISSN 1869-6929 ISSN 1869-6937 (electronic)
EAA Series
ISBN 978-3-319-02302-1 ISBN 978-3-319-02303-8 (eBook)
DOI 10.1007/978-3-319-02303-8
Springer Cham Heidelberg New York Dordrecht London

Library of Congress Control Number: 2013951154

Mathematics Subject Classification: 60G51, 60K30, 60K20, 60J75

Printed on acid-free paper

Springer is part of Springer Science+Business Media (www.springer.com)

Preface

These notes were developed whilst giving a graduate lecture course (*Nachdiplom-vorlesung*) in the fall of 2012 at the Forschungsinstitut für Mathematik (FIM), ETH Zürich, Switzerland. The same course was given in the spring of 2013 at Centro de Investigación en Matemáticas (CIMAT), Guanajuato, Mexico and, simultaneously by video-link, at Instituto de Matemáticas, UNAM in Mexico City. The title of these lecture notes may come as surprise to some readers as, to date, the term *Gerber–Shiu Risk Theory* is not widely used. One might be more tempted to simply use the title *Ruin theory for Cramér–Lundberg models* instead. However, my objective here is to focus on the recent interaction between a large body of research literature, spear-headed by Hans Gerber and Elias Shiu, concerning ever more sophisticated questions around the event of ruin for the classical Cramér–Lundberg surplus process, and the parallel evolution of the fluctuation theory of Lévy processes. The fusion of these two fields has provided economies to proofs of older results, as well as pushing classical theory much further into what one might describe as *exotic ruin theory*. The latter may be considered as the study of ruinous scenarios which involve perturbations to the surplus coming from dividend or tax payments that have a historical path dependence. These notes keep to the Cramér–Lundberg setting. However, the text has been written in a form that appeals to straightforward and accessible proofs, which take advantage, as much as possible, of the fact that Cramér–Lundberg processes have stationary and independent increments and no upward jumps.

I would like to thank Paul Embrechts for the invitation to spend six months at the FIM and the opportunity to develop and present this material. Similarly I would like to thank Maria-Emilia Caballero, Juan Carlos Pardo and Victor Rivero for the invitation to give the same course at CIMAT and UNAM simultaneously. I would also like to thank all attendees in Zürich, Guanajuato and Mexico City for their comments (especially Jean Bertoin, Leif Döring, Juan Carlos Pardo and Victor Rivero). Whilst I was in Switzerland I had the opportunity to meet Hans Gerber and Hansjörg Albrecher, with whom I had extensive discussion regarding some of the material in this book. I would also like to express my gratitude for their many observations and comments. I would also like to thank Nick Bingham and Erik Baurdoux who have diligently attended to mathematical errors, references and my use of the English language. Finally, Springer produced three anonymous referee reports whose useful comments I am also grateful for.

Guanajuato, Mexico
May, 2013

Andreas E. Kyprianou

Contents

Chapter 1
Introduction

In this brief introductory chapter, we shall outline the basic context of these lecture notes. In particular, we shall explain what we understand by so-called *Gerber–Shiu* theory and the role that it has played in classical ruin theory.

1.1 The Cramér–Lundberg Process

The beginning of ruin theory is based around a very basic model for the evolution of the wealth, or *surplus*, of an insurance company, known as the *Cramér–Lundberg process*. In the classical model, the insurance company is assumed to collect premiums at a constant rate $\mathtt{c} > 0$, whereas claims arrive successively according to the times of a Poisson process, henceforth denoted by $N = \{N_t : t \geq 0\}$, with rate $\lambda > 0$. These claims, indexed in order of appearance $\{\xi_i : i = 1, 2, \ldots\}$, are independent and identically distributed[1] with common distribution F, which is concentrated on $(0, \infty)$. The dynamics of the Cramér–Lundberg process are described by $X = \{X_t : t \geq 0\}$, where

$$X_t = \mathtt{c}t - \sum_{i=1}^{N_t} \xi_i, \quad t \geq 0, \tag{1.1}$$

is the gain between time 0 and t. Here, we use standard notation in that a sum, for example of the form $\sum_{i=1}^{0} \cdot$, is understood to be equal to zero. We assume that X is defined on a stochastic basis $(\Omega, \mathbb{F}, \mathcal{F}, \mathbb{P})$, where $\mathbb{F} := \{\mathcal{F}_t : t \geq 0\}$ is the natural filtration generated by X. When the initial surplus of our insurance company is valued at $x > 0$, we may consider the evolution of the surplus to follow the dynamics of $x + X$ under $\mathbb{P}$.

The Cramér–Lundberg process, $(X, \mathbb{P})$, is nothing but the difference of a linear trend and a compound Poisson process with positive jumps. Accordingly, it is easy to verify that it conforms to the definition of a so-called Lévy process, given below.

[1] Henceforth written i.i.d. for short.

A.E. Kyprianou, *Gerber–Shiu Risk Theory*, EAA Series,
DOI 10.1007/978-3-319-02303-8_1,

Definition 1.1 A process $X = \{X_t : t \geq 0\}$ with law $\mathbb{P}$ is said to be a *Lévy process* if it possesses the following properties:

(i) The paths of X are $\mathbb{P}$-almost surely right-continuous with left limits.
(ii) $\mathbb{P}(X_0 = 0) = 1$.
(iii) For $0 \leq s \leq t$, $X_t - X_s$ is equal in distribution to X_{t-s}.
(iv) For all $n \in \mathbb{N}$ and $0 \leq s_1 \leq t_1 \leq s_2 \leq t_2 \leq \ldots \leq s_n \leq t_n < \infty$, the increments $X_{t_i} - X_{s_i}$, $i = 1, \ldots, n$, are independent.

Whilst our computations in this text will largely remain within the confines of the Cramér–Lundberg model, we shall, as much as possible, appeal to mathematical reasoning which is handed down from the general theory of Lévy process. Specifically, our analysis will predominantly appeal to *martingale theory* as well as *excursion theory*. The latter of these two concerns the decomposition of the path of X into a sequence of sojourns from its running maximum or, indeed, from its running minimum.

Many of the arguments we give will apply, either directly or with minor modification, to the setting of general *spectrally negative Lévy processes*. These are Lévy processes which do not experience positive jumps and which do not have monotone paths. In the forthcoming chapters, we have deliberately stepped back from treating the case of general spectrally negative Lévy processes in order to keep the presentation as mathematically light as possible. Nonetheless, many of the arguments we give are robust enough to apply to the case of general spectrally negative Lévy processes, either verbatim or with minor modification. At the very end, in Chap. 9, we will spend a little time discussing the connection with the general spectrally negative setting.

As a Lévy processes, it is well understood that Cramér–Lundberg processes are strong Markov[2] and, henceforth, we shall prefer to work with the probabilities $\{\mathbb{P}_x : x \in \mathbb{R}\}$, where, thanks to spatial homogeneity, for $x \in \mathbb{R}$, $(X, \mathbb{P}_x)$ is equal in law to $x + X$ under $\mathbb{P}$. For convenience, we shall always prefer to write $\mathbb{P}$ instead of $\mathbb{P}_0$.

Recall that the random variable $\tau \in [0, \infty]$ is a *stopping time* with respect to $\mathbb{F}$ if and only if, for all $t \geq 0$, $\{\tau \leq t\} \in \mathcal{F}_t$. Moreover, to each stopping time τ, we associate the sigma-algebra

$$\mathcal{F}_\tau := \big\{A \in \mathcal{F} : A \cap \{\tau \leq t\} \in \mathcal{F}_t \text{ for all } t \geq 0\big\}.$$

(Note, it is a simple exercise to verify that $\mathcal{F}_\tau$ is a sigma-algebra.) The standard way of expressing the strong Markov property for a one-dimensional process such as X is as follows: For any Borel set B, on $\{\tau < \infty\}$,

$$\mathbb{P}(X_{\tau+s} \in B | \mathcal{F}_\tau) = \mathbb{P}\big(X_{\tau+s} \in B | \sigma(X_\tau)\big) = h(X_\tau, s),$$

[2]We assume that the reader is familiar with the basic theory of Markov processes and, in particular, the use of the (strong) Markov property.

where $h(x,s) = \mathbb{P}_x(X_s \in B)$. On account of the fact that X has stationary and independent increments, we may also state the strong Markov property in a slightly refined form.

Theorem 1.2 *Suppose that τ is a stopping time. On $\{\tau < \infty\}$, define the process $\widetilde{X} = \{\widetilde{X}_t : t \geq 0\}$ by*

$$\widetilde{X}_t = X_{\tau+t} - X_\tau, \quad t \geq 0.$$

Then, on the event $\{\tau < \infty\}$, the process $\widetilde{X}$ is independent of $\mathcal{F}_\tau$ and has the same law as X.

1.2 The Classical Problem of Ruin

Financial ruin in the Cramér–Lundberg model (or just *ruin* for short) will occur if the surplus of the insurance company drops below zero. Since this will happen with probability one if $\mathbb{P}(\liminf_{t\to\infty} X_t = -\infty) = 1$, in order to avoid this situation, it is usual to impose an additional assumption that

$$\mathbb{P}\Big(\lim_{t\to\infty} X_t = \infty\Big) = 1. \tag{1.2}$$

Write $\mu = \int_{(0,\infty)} x F(\mathrm{d}x)$ for the common mean of the i.i.d. claim sizes $\{\xi_i : i = 1, 2, \ldots\}$. A sufficient condition to guarantee (1.2) is that

$$\mathrm{c} - \lambda\mu > 0, \tag{1.3}$$

the so-called *security loading condition*. To see why this guarantees (1.2), note that the Strong Law of Large Numbers for Poisson processes, which states that $\lim_{t\to\infty} N_t/t = \lambda$ a.s., and the obvious fact that $\lim_{t\to\infty} N_t = \infty$ a.s. imply that

$$\lim_{t\to\infty} \frac{X_t}{t} = \lim_{t\to\infty} \left(\frac{x}{t} + \mathrm{c} - \frac{N_t}{t} \frac{\sum_{i=1}^{N_t} \xi_i}{N_t} \right) = \mathbb{E}(X_1) = \mathrm{c} - \lambda\mu > 0 \quad \text{a.s.}, \tag{1.4}$$

from which (1.2) follows. We shall see later that the positive security loading in (1.3) is also a necessary condition for (1.2). Note that (1.3) also implies that $\mu < \infty$.

Under the security loading condition, it follows that ruin will occur only with probability less than one. The most basic question that one can therefore ask under such circumstances is: What is the probability of ruin when the initial surplus is equal to $x > 0$? This involves giving an expression for $\mathbb{P}_x(\tau_0^- < \infty)$, where[3]

$$\tau_0^- := \inf\{t > 0 : X_t < 0\}.$$

The *Pollaczek–Khintchine formula* does just this.

[3]Throughout this text, we use the standard definition $\inf \emptyset := \infty$.

Theorem 1.3 (Pollaczek–Khintchine formula) *Suppose that $\lambda\mu/\mathrm{c} < 1$. For all $x \geq 0$,*

$$1 - \mathbb{P}_x\big(\tau_0^- < \infty\big) = (1-\rho)\sum_{k\geq 0}\rho^k \eta^{*k}(x), \tag{1.5}$$

where $\rho = \lambda\mu/\mathrm{c}$,

$$\eta(x) = \frac{1}{\mu}\int_0^x \big[1 - F(y)\big]\mathrm{d}y, \quad x \geq 0,$$

and, for $k \geq 0$, we understand η^{*k} to be the k-fold convolution of η with the special understanding that $\eta^{*0}(\mathrm{d}x) = \delta_0(\mathrm{d}x)$.

It is not our intention to dwell on the Pollaczek–Khintchine formula at this point in time, although we shall rederive it later in this text. This classical result is a cornerstone of what is known as *insurance risk theory*. Its connection to renewal theory is the inspiration behind a whole body of research literature addressing more elaborate questions concerning the ruin problem. Our aim in this text is to give an overview of the state of the art in this respect. Amongst the large number of names active in this field, one may note, in particular, the many original and varied contributions of Hans Gerber and Elias Shiu. In recognition of their foundational work spanning several decades, we accordingly refer to the collective results that we present here as *Gerber–Shiu risk theory*.

1.3 Gerber–Shiu Expected Discounted Penalty Functions

Following Theorem 1.3, an obvious direction in which to turn one's attention is to look at the joint distribution of τ_0^-, $-X_{\tau_0^-}$ and $X_{\tau_0^- -}$. These three quantities can otherwise be called the *time of ruin*, the *deficit at ruin* and *the surplus prior to ruin*. In their well-cited paper of 1998,[4] Gerber and Shiu introduce the so-called *expected discounted penalty function* as follows. Suppose that $f : (0,\infty)^2 \to [0,\infty)$ is any bounded, measurable function. Then the associated expected discounted penalty function with force of interest $q \geq 0$, when the initial surplus is equal to $x \geq 0$, is given by

$$\mathrm{GS}_f(x,q) := \mathbb{E}_x\big[\mathrm{e}^{-q\tau_0^-} f(-X_{\tau_0^-}, X_{\tau_0^- -})\mathbf{1}_{(\tau_0^- < \infty)}\big].$$

Ultimately, we are interested in what we call here the *Gerber–Shiu measure*, namely the exponentially discounted joint law of the pair $(-X_{\tau_0^-}, X_{\tau_0^- -})$, denoted by

$$K^{(q)}(x, \mathrm{d}y, \mathrm{d}z) := \mathbb{E}_x\big[\mathrm{e}^{-q\tau_0^-}; -X_{\tau_0^-} \in \mathrm{d}y, X_{\tau_0^- -} \in \mathrm{d}z\big], \quad x, y, z \geq 0. \tag{1.6}$$

[4] See the historical remarks at the end of this chapter.

For the Gerber–Shiu measure, one notes the simple relation

$$\mathrm{GS}_f(x,q) = \int_{[0,\infty)} \int_{[0,\infty)} f(y,z) K^{(q)}(x,\mathrm{d}y,\mathrm{d}z).$$

The expected discounted penalty function is a well-studied object and there are many different ways to develop the expression on the right-hand side of (1.6). We shall show later in this text how the Gerber–Shiu measure can be written in terms of so-called *scale functions*. Scale functions are a natural family of functions with which one may develop many of the identities related to the event of ruin, which concern the way in which ruin occurs in a variety of different scenarios. We shall spend quite some time discussing the recent theory of scale functions later on in this text.

1.4 Exotic Gerber–Shiu Theory

Again inspired by foundational work of Gerber and Shiu, and indeed many others, we shall also look at variants of the classical ruin problem in the setting that the Cramér–Lundberg process undergoes perturbations in its trajectory. These perturbations will represent pay-outs corresponding to dividend or taxation payments. Three specific cases that will interest us are the following.

Reflection Strategies An adaptation of the classical ruin problem, introduced by Bruno de Finetti in 1957, is to consider the payment of dividends from the surplus process to (hypothetical) shareholders. Naturally, for a given stream of dividend payments, this will continuously change the aggregate value of the surplus process and the problem of ruin will look quite different. Indeed, the event of ruin will be highly dependent on the choice of dividend payments. One may formulate the problem of finding an optimal way of paying out dividends such as to maximise the expected present value of the total income of the shareholders, under force of interest $q \geq 0$, from time zero until ruin. The optimisation is made over an appropriate class of dividend strategies. Mathematically speaking, *de Finetti's dividend problem* amounts to solving a control problem which we reproduce here.

Let $\xi = \{\xi_t : t \geq 0\}$, with $\xi_0 = 0$, be a dividend strategy consisting of a left-continuous, non-negative, non-decreasing process which is $\mathbb{F}$-adapted. The quantity ξ_t represents the cumulative dividends paid out up to time $t \geq 0$ by the insurance company whose surplus is modelled by X. The aggregate, or *controlled*, value of the surplus process, when taking account of the dividend strategy ξ, is thus $U^\xi = \{U_t^\xi : t \geq 0\}$, where $U_t^\xi = X_t - \xi_t$, $t \geq 0$. An additional constraint on ξ is that $\xi_{t+} - \xi_t \leq \max\{U_t^\xi, 0\}$ for $t \geq 0$ (i.e. lump sum dividend payments are always smaller than the available reserves).

Let Ξ be the family of dividend strategies as outlined in the previous paragraph and, for each $\xi \in \Xi$, write $\sigma^\xi = \inf\{t > 0 : U_t^\xi < 0\}$ for the time at which ruin occurs for the controlled risk process. The expected present value, with discounting at rate $q \geq 0$, associated to the dividend policy ξ, when the risk process has initial capital $x \geq 0$, is given by

$$v_\xi(x) = \mathbb{E}_x\left(\int_0^{\sigma^\xi} \mathrm{e}^{-qt}\mathrm{d}\xi_t\right).$$

De Finetti's dividend problem consists of solving the stochastic control problem

$$v^*(x) := \sup_{\xi\in\Xi} v_\xi(x), \quad x \geq 0. \tag{1.7}$$

That is, if it exists, one seeks to establish a strategy, $\xi^* \in \Xi$, such that $v^* = v_{\xi^*}$.

We shall refrain from giving a complete account of this problem, other than to say that under certain conditions on the jump distribution of the Cramér–Lundberg process, X, the optimal strategy consists of a so-called *reflection strategy*, also known as a *barrier strategy*. Specifically, there exists an $\mathtt{a} \in [0, \infty)$ such that

$$\xi_t^* = (\mathtt{a} \vee \overline{X}_t) - \mathtt{a}, \quad t > 0,$$

where

$$\overline{X}_t = \sup_{s\leq t} X_s$$

is the running supremum of the surplus process. In that case, the ξ^*-controlled risk process is identical, on $t > 0$, to the process $\{\mathtt{a} - Y_t : t \geq 0\}$ under $\mathbb{P}_x$, where

$$Y_t = (\mathtt{a} \vee \overline{X}_t) - X_t, \quad t \geq 0.$$

Refraction Strategies An adaptation of the optimal control problem deals with the case that optimality is sought in a subclass of Ξ, say Ξ_δ. Specifically, Ξ_δ denotes the set of dividend strategies $\xi \in \Xi$ such that

$$\xi_t = \int_0^t \alpha_s \mathrm{d}s, \quad t \geq 0,$$

where $\{\alpha_t : t \geq 0\}$ is uniformly bounded by some constant, say $\delta > 0$. In other words, Ξ_δ consists of dividend strategies which are absolutely continuous, with uniformly bounded density.

Again, we refrain from going into the details of the solution to (1.7), other than to say that, under certain conditions, the optimal strategy, $\xi^\delta = \{\xi_t^\delta : t \geq 0\}$ in Ξ_δ turns out to satisfy

$$\xi_t^\delta = \delta \int_0^t \mathbf{1}_{(Z_s > \mathtt{b})}\mathrm{d}s, \quad t \geq 0,$$

for some $\mathtt{b} \geq 0$, where $Z = \{Z_t : t \geq 0\}$ is the controlled surplus process $X - \xi^\delta$. Each one of the pair (Z, ξ^δ) must be expressed in terms of the other and we are forced to work with the stochastic differential equation

$$\mathrm{d}Z_t = \mathrm{d}X_t - \delta \mathbf{1}_{(Z_t > \mathtt{b})}\mathrm{d}t, \quad t \geq 0.$$

This is also written in integral form as

$$Z_t = X_t - \delta \int_0^t \mathbf{1}_{(Z_s > \mathtt{b})}\mathrm{d}s, \quad t \geq 0. \tag{1.8}$$

For reasons that we shall elaborate on later, the process in (1.8) is called a *refracted process*.

Perturbation-at-Maximum Strategies Another way of perturbing the path of our Cramér–Lundberg process is by forcing payments from the surplus at times that it attains a new maximum. This may be interpreted, for example, as tax payments. To this end, consider the process

$$U_t = X_t - \int_{(0,t]} \gamma(\overline{X}_u)\mathrm{d}\overline{X}_u, \quad t \geq 0, \tag{1.9}$$

where $\gamma : [0, \infty) \to [0, \infty)$ satisfies appropriate conditions.

We distinguish two regimes, *light-perturbation* and *heavy-perturbation* regimes. The first corresponds to the case that $\gamma : [0, \infty) \to [0, 1)$ and the second to the case that $\gamma : [0, \infty) \to (1, \infty)$. The light-perturbation regime has a similar flavour to paying dividends at a weaker rate than a reflection strategy, and may be seen as a taxation on new levels of wealth. In contrast, the heavy-perturbation regime is equivalent to paying dividends at a much stronger rate than a reflection strategy. (The connection with tax payments is arguably lost.)

For each of the three scenarios described above, reflection, refraction and perturbation-at-maximum, questions concerning the way in which ruin occurs remain just as pertinent as for the case of the Cramér–Lundberg process alone. In addition, we are also interested in the distribution of the present value of payments made out of the surplus process until ruin. For example, in the case of a reflection strategy with barrier $\mathtt{a} \geq 0$ and force of interest equal to $q \geq 0$, this boils down to understanding the distribution of

$$\int_0^{\sigma_{\mathtt{a}}} \mathrm{e}^{-qt}\mathrm{d}\overline{X}_t,$$

where

$$\sigma_{\mathtt{a}} = \inf\{t > 0 : \overline{X}_t - X_t > \mathtt{a}\}.$$

1.5 Comments

For the standard theory of stochastic processes, including the basic theory of Markov processes and stopping times, see for example Kallenberg (2002). For more on the classical derivation of the standard model of a surplus process, see Lundberg (1903), Cramér (1994a, 1994b). Asmussen and Albrecher (2010) serves as an encyclopaedic reference for all matters concerning ruin theory. See also, for example, the books of Embrechts et al. (1997) and Dickson (1999), to name but a few standard texts on the classical theory. A classical derivation of the Pollaczek–Khintchine formula (also known in the actuarial literature as *Beekman's convolution formula*) can be found in Chapter XII of Feller (1971).

Within the setting of the classical Cramér–Lundberg model, Gerber and Shiu (1998) introduced the expected discounted penalty function. See also Gerber and Shiu (1997). It has been widely studied since, with too many references to list here. The special issue in volume 46, 2010 of the journal *Insurance: Mathematics and Economics* contains a selection of papers focused on the Gerber–Shiu expected discounted penalty function, with many further references therein.

An adaptation of the classical ruin problem was introduced within the setting of a discrete-time surplus process by de Finetti (1957). There, dividends are paid out to shareholders up to the moment of ruin, resulting in a discrete-time analogue of the control problem (1.7). This control problem was considered in the framework of Cramér–Lundberg processes by Gerber (1969, 1972) and then, after a long gap, by Azcue and Muler (2005). Schmidli (2008) gives an extensive account of (1.7) and variants thereof. A string of articles, each one improving on its predecessor, develops the solution to (1.7) in the setting that the surplus process is modelled by a spectrally negative Lévy process; see Avram et al. (2007), Loeffen (2008), Kyprianou et al. (2010) and Loeffen and Renaud (2010). The variant of (1.7) resulting in refraction strategies was studied by Jeanblanc and Shiryaev (1995) and Asmussen and Taksar (1997) in the diffusive setting, Gerber and Shiu (2006b) in the Cramér–Lundberg setting and Kyprianou et al. (2012) in the setting of spectrally negative Lévy processes.

In the setting of the classical Cramér–Lundberg model, Albrecher and Hipp (2007) introduced the idea of tax payments as in (1.9), for the case that γ is a constant in $(0, 1)$. This model was quickly generalised by Albrecher et al. (2008, 2011), Kyprianou and Zhou (2009) and Kyprianou and Ott (2012).

Chapter 2
The Wald Martingale and the Maximum

In this chapter, we shall introduce the first of our two key martingales and consider two immediate applications. In the first application, we will use the martingale to construct a change of measure with respect to $\mathbb{P}$ and thereby consider the dynamics of X under the new law. In the second application, we shall use the martingale to study the law of the process $\overline{X} = \{\overline{X}_t : t \geq 0\}$, where we recall that

$$\overline{X}_t = \sup_{s \leq t} X_s, \quad t \geq 0. \tag{2.1}$$

In particular, we shall discover that the position of the trajectory of $\overline{X}$, when sampled at an independent and exponentially distributed time, is again exponentially distributed.

2.1 Laplace Exponent

A key quantity in the forthcoming analysis is the *Laplace exponent* of the Cramér–Lundberg process, whose definition is contained in the following lemma.

Lemma 2.1 *For all $\theta \geq 0$ and $t \geq 0$, we have*

$$\mathbb{E}\big(\mathrm{e}^{\theta X_t}\big) = \exp\big\{\psi(\theta)t\big\},$$

where the Laplace exponent ψ satisfies

$$\psi(\theta) := c\theta - \lambda \int_{(0,\infty)} \big(1 - \mathrm{e}^{-\theta x}\big) F(\mathrm{d}x). \tag{2.2}$$

Proof Given the definition (1.1), one easily sees that it suffices to prove that

$$\mathbb{E}\big(\mathrm{e}^{-\theta \sum_{i=1}^{N_t} \xi_i}\big) = \exp\left\{-\lambda t \int_{(0,\infty)} \big(1 - \mathrm{e}^{-\theta x}\big) F(\mathrm{d}x)\right\}, \tag{2.3}$$

A.E. Kyprianou, *Gerber–Shiu Risk Theory*, EAA Series,
DOI 10.1007/978-3-319-02303-8_2,

for $\theta, t \geq 0$. To establish (2.3), we make use of the fact that N_t is independent of $\{\xi_i : i \geq 1\}$ and Poisson distributed with rate λt. More precisely,

$$\begin{aligned}\mathbb{E}(\mathrm{e}^{-\theta \sum_{i=1}^{N_t} \xi_i}) &= \sum_{n=0}^{\infty} \mathbb{E}(\mathrm{e}^{-\theta \sum_{i=1}^{n} \xi_i})\mathrm{e}^{-\lambda t}\frac{(\lambda t)^n}{n!} \\ &= \sum_{n=0}^{\infty} \left[\mathbb{E}(\mathrm{e}^{-\theta \xi_1})\right]^n \mathrm{e}^{-\lambda t}\frac{(\lambda t)^n}{n!} \\ &= \exp\left\{-\lambda t\left(1 - \mathbb{E}(\mathrm{e}^{-\theta \xi_1})\right)\right\} \\ &= \exp\left\{-\lambda t \int_{(0,\infty)} (1 - \mathrm{e}^{-\theta x}) F(\mathrm{d}x)\right\},\end{aligned}$$

for all $\theta, t \geq 0$. □

As we shall see, the Laplace exponent (2.2) is used as a way of identifying certain characteristics of Cramér–Lundberg processes. To this end, let us start by looking at the shape of (2.2). Straightforward differentiation, with the help of the Dominated Convergence Theorem, tells us that, for all $\theta > 0$,

$$\psi''(\theta) = \lambda \int_{(0,\infty)} x^2 \mathrm{e}^{-\theta x} F(\mathrm{d}x) > 0,$$

which in turn implies that ψ is strictly convex on $(0, \infty)$. Integration by parts allows us to write

$$\psi(\theta) = \mathrm{c}\theta - \lambda\theta \int_{(0,\infty)} \mathrm{e}^{-\theta x}\overline{F}(x)\mathrm{d}x, \quad \theta \geq 0, \tag{2.4}$$

where $\overline{F}(x) := 1 - F(x)$, $x \geq 0$. Moreover, this representation allows us to deduce that

$$\lim_{\theta \to \infty} \frac{\psi(\theta)}{\theta} = \mathrm{c}$$

and

$$\psi'(0+) = \lim_{\theta \to 0} \frac{\psi(\theta)}{\theta} = \mathrm{c} - \lambda\mu = \mathbb{E}(X_1),$$

where the left-hand side is the right derivative of ψ at the origin, the final equality follows from (1.4) and we recall that $\mu := \int_{(0,\infty)} x F(\mathrm{d}x) \in (0, \infty]$. The security loading condition (1.3) can thus be alternatively expressed simply as $\psi'(0+) > 0$.

A quantity which will also repeatedly appear in our computations is the right inverse of ψ. That is,

$$\Phi(q) := \sup\{\theta \geq 0 : \psi(\theta) = q\}, \tag{2.5}$$

for $q \geq 0$. Thanks to the strict convexity of ψ and that $\lim_{\theta \to \infty} \psi(\theta) = \infty$, we can say that there is exactly one solution in $[0, \infty)$ to the equation $\psi(\theta) = q$, when $q > 0$, and at most two when $q = 0$. The number of solutions in the latter of these

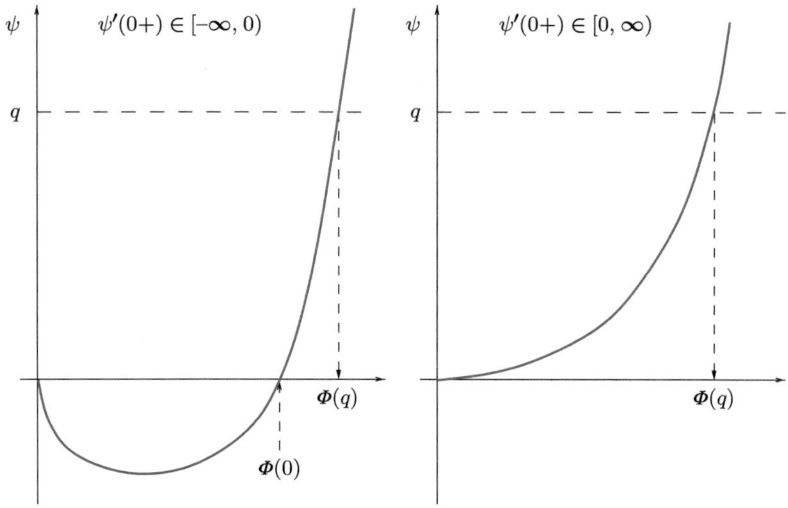

Fig. 2.1 Two examples of ψ, the Laplace exponent of a Cramér–Lundberg process, corresponding to the cases $\psi'(0+) < 0$ and $\psi'(0+) \geq 0$, respectively

two cases depends on the value of $\psi'(0+)$. Indeed, when $\psi'(0+) \geq 0$, then $\theta = 0$ is the only solution to $\psi(\theta) = 0$. When $\psi'(0+) < 0$, there are two solutions, one at $\theta = 0$ and a second solution, in $(0, \infty)$, which, by definition, gives the value of $\Phi(0)$; see Fig. 2.1.

2.2 First Exponential Martingale

For each $\beta > 0$, define the process $\mathcal{E}(\beta) = \{\mathcal{E}_t(\beta) : t \geq 0\}$ by

$$\mathcal{E}_t(\beta) := \mathrm{e}^{\beta X_t - \psi(\beta)t}, \quad t \geq 0. \tag{2.6}$$

Theorem 2.2 *Fix $\beta > 0$. The process $\mathcal{E}(\beta)$ is a $\mathbb{P}$-martingale with respect to $\mathbb{F}$.*

Proof Note that the process $\mathcal{E}(\beta)$ is $\mathbb{F}$-adapted. With this in hand, it suffices to check that, for all $\beta > 0$ and $s, t \geq 0$, $\mathbb{E}[\mathcal{E}_{t+s}(\beta)|\mathcal{F}_t] = \mathcal{E}_t(\beta)$. On account of positivity, this would immediately show that $\mathbb{E}[|\mathcal{E}_t(\beta)|] < \infty$, for all $t \geq 0$, which is also required for $\mathcal{E}(\beta)$ to be a martingale.

Thanks to stationary and independent increments, $\mathcal{F}$-adaptedness as well as Lemma 2.1, for all $\beta, s, t \geq 0$,

$$\begin{aligned}\mathbb{E}\big[\mathcal{E}_{t+s}(\beta)|\mathcal{F}_t\big] &= \mathcal{E}_t(\beta)\mathbb{E}\big[\mathrm{e}^{\beta(X_{t+s}-X_t)-\psi(\beta)s}|\mathcal{F}_t\big] \\ &= \mathcal{E}_t(\beta)\mathbb{E}\big[\mathrm{e}^{\beta X_s}\big]\mathrm{e}^{-\psi(\beta)s} \\ &= \mathcal{E}_t(\beta)\end{aligned}$$

and the proof is complete. □

This martingale is known as the *Wald martingale*. See Sect. 2.5 for further historical details.

2.3 Esscher Transform

Fix $\beta > 0$ and $x \in \mathbb{R}$. Normalising $\mathcal{E}(\beta)$ by its expectation, we may use the resulting mean-one martingale to perform a change of measure on $(X, \mathbb{P}_x)$ via

$$\left.\frac{\mathrm{d}\mathbb{P}_x^\beta}{\mathrm{d}\mathbb{P}_x}\right|_{\mathcal{F}_t} = \frac{\mathcal{E}_t(\beta)}{\mathcal{E}_0(\beta)} = \mathrm{e}^{\beta(X_t - x) - \psi(\beta)t}, \quad t \geq 0. \tag{2.7}$$

In the special case that $x = 0$, we shall write $\mathbb{P}^\beta$ in place of $\mathbb{P}_0^\beta$. Since the process X under $\mathbb{P}_x$ may be written as $x + X$ under $\mathbb{P}$, it is not difficult to see that the change of measure on $(X, \mathbb{P}_x)$ corresponds to the analogous change of measure on $(X, \mathbb{P})$. Also known as the *Esscher transform*, (2.7) alters the law of X. It is related to the Esscher transform for random variables. For example, for the distribution F, its Esscher transform is the distribution

$$F_\beta(\mathrm{d}x) := \frac{\mathrm{e}^{-\beta x}}{m(\beta)} F(\mathrm{d}x), \quad x > 0,$$

for some $\beta > 0$, where $m(\beta) = \int_{(0,\infty)} \mathrm{e}^{-\beta x} F(\mathrm{d}x)$. For the forthcoming computations, it is important that we understand the dynamics of X under $\mathbb{P}^\beta$.

Theorem 2.3 *Fix $\beta > 0$. The process $(X, \mathbb{P}^\beta)$ is equal in law to a Cramér–Lundberg process with premium rate $\mathtt{c}$ and claims that arrive at rate $\lambda m(\beta)$ with common distribution F_β. Said another way, the process $(X, \mathbb{P}^\beta)$ is equal in law to X^β, where $X^\beta := \{X_t^\beta : t \geq 0\}$ is a Cramér–Lundberg process with Laplace exponent*

$$\psi_\beta(\theta) := \psi(\theta + \beta) - \psi(\beta), \quad \theta \geq 0.$$

Proof For all $0 \leq s \leq t \leq u < \infty$, $\theta \geq 0$ and $A \in \mathcal{F}_s$, with the help of stationary and independent increments of $(X, \mathbb{P})$, we have that

$$\begin{aligned}
\mathbb{E}^\beta\big[\mathbf{1}_A \mathrm{e}^{\theta(X_u - X_t)}\big] &= \mathbb{E}\big[\mathbf{1}_A \mathrm{e}^{\beta X_t - \psi(\beta)t} \mathrm{e}^{(\theta+\beta)(X_u - X_t)}\big] \mathrm{e}^{-\psi(\beta)(u-t)} \\
&= \mathbb{E}\big[\mathbf{1}_A \mathrm{e}^{\beta X_t - \psi(\beta)t}\big] \mathbb{E}\big[\mathrm{e}^{(\theta+\beta)X_{u-t}}\big] \mathrm{e}^{-\psi(\beta)(u-t)} \\
&= \mathbb{E}\big[\mathbf{1}_A \mathrm{e}^{\beta X_s - \psi(\beta)s}\big] \mathbb{E}\mathrm{e}^{\psi(\beta+\theta)(u-t)} \mathrm{e}^{-\psi(\beta)(u-t)} \\
&= \mathbb{P}^\beta(A) \mathrm{e}^{\psi_\beta(\theta)(u-t)},
\end{aligned} \tag{2.8}$$

where in the second equality we have conditioned on $\mathcal{F}_t$ and in the third equality we have conditioned on $\mathcal{F}_s$ and used the martingale property of $\mathcal{E}(\beta)$. It now follows from (2.8) that, for all $0 \le v \le s \le t \le u < \infty$ and $\theta_1, \theta_2 \ge 0$,

$$\mathbb{E}^\beta\left[\mathrm{e}^{\theta_1(X_s - X_v)}\mathrm{e}^{\theta_2(X_u - X_t)}\right] = \mathrm{e}^{\psi_\beta(\theta_1)(s-v)+\psi_\beta(\theta_2)(u-t)}.$$

Using a straightforward argument by induction, again using (2.8), we also have that, for all $n \in \mathbb{N}$, $0 \le s_1 \le t_1 \le \ldots \le s_n \le t_n < \infty$ and $\theta_1, \ldots, \theta_n \ge 0$,

$$\mathbb{E}^\beta\left[\prod_{j=1}^n \mathrm{e}^{\theta_j(X_{t_j} - X_{s_j})}\right] = \prod_{j=1}^n \mathrm{e}^{\psi_\beta(\theta_j)(t_j - s_j)}. \tag{2.9}$$

Moreover, a brief computation shows that

$$\psi_\beta(\theta) = \mathrm{c}\theta - \lambda m(\beta)\int_{(0,\infty)} (1 - \mathrm{e}^{-\theta x})\frac{\mathrm{e}^{-\beta x}}{m(\beta)}F(\mathrm{d}x), \quad \theta \ge 0.$$

Coupled with (2.9), this shows that $(X, \mathbb{P}^\beta)$ has stationary and independent increments which are equal in law to those of a Cramér–Lundberg process with premium rate c, arrival rate of claims $\lambda m(\beta)$ and distribution of claims $\mathrm{e}^{-\beta x}F(\mathrm{d}x)/m(\beta)$. Since the measures $\mathbb{P}^\beta$ and $\mathbb{P}$ are equivalent on $\mathcal{F}_t$, for all $t \ge 0$, then the property that X has paths that are almost surely right-continuous with left limits and no positive jumps on $[0, t]$ carries over to the measure $\mathbb{P}^\beta$. □

The Esscher transform may also be formulated at stopping times.

Corollary 2.4 *Under the conditions of Theorem* 2.3, *if τ is an $\mathbb{F}$-stopping time, then*

$$\left.\frac{\mathrm{d}\mathbb{P}^\beta}{\mathrm{d}\mathbb{P}}\right|_{\mathcal{F}_\tau} = \mathcal{E}_\tau(\beta) \quad \text{on } \{\tau < \infty\}.$$

Said another way, for all $A \in \mathcal{F}_\tau$, we have

$$\mathbb{P}^\beta(A, \tau < \infty) = \mathbb{E}\big(\mathbf{1}_{(A,\tau<\infty)}\mathcal{E}_\tau(\beta)\big).$$

Proof By definition, if $A \in \mathcal{F}_\tau$, then $A \cap \{\tau \le t\} \in \mathcal{F}_t$. Hence

$$\begin{aligned}\mathbb{P}^\beta\big(A \cap \{\tau \le t\}\big) &= \mathbb{E}\big(\mathcal{E}_t(\beta)\mathbf{1}_{(A,\tau\le t)}\big)\\ &= \mathbb{E}\big(\mathbf{1}_{(A,\tau\le t)}\mathbb{E}\big(\mathcal{E}_t(\beta)|\mathcal{F}_\tau\big)\big)\\ &= \mathbb{E}\big(\mathcal{E}_\tau(\beta)\mathbf{1}_{(A,\tau\le t)}\big),\end{aligned}$$

where in the third equality we have used the strong Markov property as well as the martingale property for $\mathcal{E}(\beta)$. Now taking limits as $t \to \infty$, the result follows with the help of the Monotone Convergence Theorem. □

2.4 Distribution of the Maximum

We want to use the Esscher transform to characterise the law of the first passage times

$$\tau_x^+ := \inf\{t > 0 : X_t > x\},$$

for $x \geq 0$, and subsequently the law of the running maximum when sampled at an independent and exponentially distributed time. Note that the stopping time τ_x^+ may be infinite in value, depending on the long-term behaviour of the process X. Accordingly, in the theorem below, where τ_x^+ appears in an exponent, we will understand $\mathrm{e}^{-\infty} := 0$.

Theorem 2.5 *For $x \geq 0$ and $q > 0$,*

$$\mathbb{E}\big(\mathrm{e}^{-q\tau_x^+}\big) = \mathrm{e}^{-\Phi(q)x},$$

where we recall that $\Phi(q)$ is given by (2.5). By taking limits as $q \to 0$, it also follows that

$$\mathbb{P}\big(\tau_x^+ < \infty\big) = \mathrm{e}^{-\Phi(0)x},$$

for $x \geq 0$.

Proof Using the fact that X has no positive jumps, it must follow that $X_{\tau_x^+} = x$ on $\{\tau_x^+ < \infty\}$. With the help of the strong Markov property we have that

$$\begin{aligned}
&\mathbb{E}\big(\mathrm{e}^{\Phi(q)X_t - qt} | \mathcal{F}_{\tau_x^+}\big) \\
&= \mathbf{1}_{(\tau_x^+ \geq t)} \mathrm{e}^{\Phi(q)X_t - qt} + \mathbf{1}_{(\tau_x^+ < t)} \mathrm{e}^{\Phi(q)x - q\tau_x^+} \mathbb{E}\big(\mathrm{e}^{\Phi(q)(X_t - X_{\tau_x^+}) - q(t - \tau_x^+)} | \mathcal{F}_{\tau_x^+}\big) \\
&= \mathrm{e}^{\Phi(q)X_{t\wedge\tau_x^+} - q(t\wedge\tau_x^+)}, \qquad (2.10)
\end{aligned}$$

where, in the final equality, we have used the fact that $\mathbb{E}(\mathcal{E}_t(\Phi(q))) = 1$ for all $t \geq 0$. Using this fact again together with the law of total probability, we get, by taking expectations again in (2.10),

$$\mathbb{E}\big(\mathrm{e}^{\Phi(q)X_{t\wedge\tau_x^+} - q(t\wedge\tau_x^+)}\big) = 1.$$

Noting that the expression in the latter expectation is bounded above by $\mathrm{e}^{\Phi(q)x}$, an application of dominated convergence yields

$$\mathbb{E}\big(\mathrm{e}^{\Phi(q)x - q\tau_x^+}\big) = 1,$$

which is equivalent to the statement of the theorem. □

We recover the promised distributional information about the maximum process (2.1) in the next corollary. In its statement, we understand an exponential random variable with rate 0 to be infinite in value with probability one.

Corollary 2.6 *Fix $q \geq 0$ and let $\mathbf{e}_q$ be an exponentially distributed random variable with rate q, which is independent of X. Then $\overline{X}_{\mathbf{e}_q}$ is exponentially distributed with parameter $\Phi(q)$.*

Proof First suppose that $q > 0$. The result is an easy consequence of the fact that

$$\mathbb{P}(\overline{X}_{\mathbf{e}_q} > x) = \mathbb{P}(\tau_x^+ < \mathbf{e}_q) = \mathbb{E}\left(\int_0^\infty q\mathrm{e}^{-qt}\mathbf{1}_{(\tau_x^+ < t)}\right) = \mathbb{E}(\mathrm{e}^{-q\tau_x^+}),$$

together with the conclusion of Theorem 2.5. For the remaining case that $q = 0$, note with the help of the last part of Theorem 2.5 that

$$\mathbb{P}(\overline{X}_\infty > x) = \mathbb{P}(\tau_x^+ < \infty) = \mathrm{e}^{-\Phi(0)x},$$

and the proof is complete. □

2.5 Comments

The idea of *tilting* a distribution by exponentially weighting its probability distribution function was introduced by Esscher (1932). This idea lends itself well to changes of measure in the theory of stochastic processes, in particular for Lévy processes. The Wald martingale can be traced back to Wald (1944, 1945). The associated Esscher transform is analogous to the exponential martingale for Brownian motion and the role that it plays in the classical Cameron–Martin–Girsanov change of measure. Indeed, the theory presented here may be extended to the general class of spectrally negative Lévy processes, which includes Cramér–Lundberg processes and Brownian motion. See for example Chap. 3 of Kyprianou (2013). The Esscher transform plays a prominent role in mathematical finance as well as insurance mathematics; see for example the discussion in the paper of Gerber and Shiu (1994) and references therein. The style of reasoning in the proof of Theorem 2.5 is inspired by the classical computations of Wald (1944) for random walks, see also Bingham (1975) and Gerber (1990).

Chapter 3
The Kella–Whitt Martingale and the Minimum

We move now to the second of our two key martingales. In a similar spirit to the previous chapter, we shall use the martingale to study the law of the process $\underline{X} = \{\underline{X}_t : t \geq 0\}$, where

$$\underline{X}_t := \inf_{s \leq t} X_s, \quad t \geq 0. \tag{3.1}$$

As with the case of $\overline{X}$, we shall characterise the law of $\underline{X}$ when sampled at an independent and exponentially distributed time. Unlike the case of $\overline{X}$ however, this will not turn out be exponentially distributed. In order to carry out the necessary analysis, we will need to pass through two sections of preparatory material.

3.1 The Cramér–Lundberg Process Reflected in Its Supremum

Fix $x \geq 0$. Define the process $Y^x = \{Y_t^x : t \geq 0\}$, where

$$Y_t^x := (x \vee \overline{X}_t) - X_t, \quad t \geq 0.$$

Lemma 3.1 *For each $x \geq 0$, Y^x is a Markov process.*

Proof Let $\widetilde{X}_s = X_{t+s} - X_t$, $s, t \geq 0$, and recall that $\{\widetilde{X}_s : s \geq 0\}$ is independent of $\mathcal{F}_t$ with the same law as $(X, \mathbb{P})$. Note that, for $t, s \geq 0$,

$$\begin{aligned}
(x \vee \overline{X}_{t+s}) - X_{t+s} &= \Big(x \vee \overline{X}_t \vee \sup_{u \in [t,t+s]} X_u\Big) - X_t - \widetilde{X}_s \\
&= \Big[(x \vee \overline{X}_t - X_t) \vee \Big(\sup_{u \in [t,t+s]} X_u - X_t\Big)\Big] - \widetilde{X}_s \\
&= \Big[Y_t^x \vee \sup_{u \in [0,s]} \widetilde{X}_u\Big] - \widetilde{X}_s.
\end{aligned}$$

From the right-hand side above, it is clear that the law of Y_{t+s}^x depends only on $\{Y_u^x : u \leq t\}$ through the value of Y_t^x. Hence $\{Y_t^x : t \geq 0\}$ is a Markov process. □

A.E. Kyprianou, *Gerber–Shiu Risk Theory*, EAA Series,
DOI 10.1007/978-3-319-02303-8_3,

Remark 3.2 Note that the argument given in the proof above shows that, if τ is any stopping time with respect to $\mathbb{F}$, then, on $\{\tau < \infty\}$,

$$(x \vee \overline{X}_{\tau+s}) - X_{\tau+s} = \left[Y^x_\tau \vee \sup_{u\in[0,s]} \widetilde{X}_u\right] - \widetilde{X}_s,$$

where now, $\widetilde{X}_s := X_{\tau+s} - X_\tau$, $s \geq 0$, is independent of $\mathcal{F}_\tau$ and has the same law as $(X, \mathbb{P})$. In other words,

$$Y^x_{\tau+s} = \widetilde{Y}^z_s \quad \text{such that} \quad z = Y^x_\tau,$$

where $\widetilde{Y}^z$ is independent of $\mathcal{F}_\tau$ and equal in law to Y^z.

Remark 3.3 It is also possible to argue, in the style of the proof of Lemma 3.1, that, for each $y \leq 0$, the process Z^y, given by

$$Z^y_t := X_t - (y \wedge \underline{X}_t), \quad t \geq 0,$$

is also a Markov process. In fact, one can go much further and show that, for $x \geq 0$, $y \leq 0$, the quadruplet (Y^x, Z^y, X, N) is also Markovian. Specifically, for each $s, t \geq 0$,

$$\left(Y^x_{t+s}, Z^y_{t+s}, X_{t+s}, N_{t+s}\right) = \left(\widetilde{Y}^u_s, \widetilde{Z}^v_s, w + \widetilde{X}_s, n + \widetilde{N}_s\right)$$

such that $u = Y^x_t$, $v = Z^y_t$, $w = X_t$ and $n = N_t$, where $\{(\widetilde{Y}^u_s, \widetilde{Z}^v_s, \widetilde{X}_s, \widetilde{N}_s) : s \geq 0\}$ is independent of $\mathcal{F}_t$ and equal in law to (Y^u, Z^v, X, N) under $\mathbb{P}$. Again, one also easily replaces t by an $\mathbb{F}$-stopping time in the above observation, as in the previous remark.

3.2 A Useful Poisson Integral

In the next section, we will come across some functionals of the driving Poisson process $N = \{N_t : t \geq 0\}$. Specifically, we will be interested in expected sums of the form

$$\mathbb{E}\left[\sum_{i=1}^{N_t} f\left(Y^x_{T_i-}, \xi_i\right)\right], \quad x, t \geq 0,$$

where $f : [0, \infty) \times (0, \infty) \to [0, \infty)$ is measurable, $\{T_i : i \geq 1\}$ are the arrival times in the process N and recall that $\{\xi_i : i \geq 1\}$ are the i.i.d. subsequent claim sizes of X with common distribution F. We need the following result.

Theorem 3.4 (Compensation formula) *For all non-negative, bounded, measurable f and $x, t \geq 0$,*

$$\mathbb{E}\left[\sum_{i=1}^{N_t} f\left(Y^x_{T_i-}, \xi_i\right)\right] = \lambda \int_0^t \int_{(0,\infty)} \mathbb{E}\left[f\left(Y^x_s, u\right)\right] F(\mathrm{d}u)\mathrm{d}s. \tag{3.2}$$

Proof With the help of Fubini's Theorem, we can write

$$\mathbb{E}\left[\sum_{i=1}^{\infty}\mathbf{1}_{(T_i\le t)}f\left(Y^x_{T_i-},\xi_i\right)\right]=\sum_{i=1}^{\infty}\mathbb{E}\left[\mathbf{1}_{(T_i\le t)}f\left(Y^x_{T_i-},\xi_i\right)\right]. \tag{3.3}$$

Note that $T_i=\inf\{t>0:N_t=i\}$ and hence each T_i is a stopping time. Note also that, for each $i\ge 1$, the terms $f(Y^x_{T_i-})$ are each measurable in the sigma-algebra $\mathcal{H}_i:=\sigma(\{N_s:s\le T_i\},\{\xi_j:j=1,\ldots,i-1\})$. For each of the expectations in the sum on the right-hand side of (3.3), by first conditioning on $\mathcal{H}_i$ and then applying Fubini's Theorem again, it follows that

$$\sum_{i=1}^{\infty}\int_{(0,\infty)}\mathbb{E}\left[\mathbf{1}_{(T_i\le t)}f\left(Y^x_{T_i-},u\right)\right]F(\mathrm{d}u)=\int_{(0,\infty)}\mathbb{E}\left[\sum_{i=1}^{N_t}f\left(Y^x_{T_i-},u\right)\right]F(\mathrm{d}u).$$

The proof is therefore complete as soon as we show that, for all $x,t\ge 0$ and $u>0$,

$$\mathbb{E}\left[\sum_{i=1}^{N_t}f\left(Y^x_{T_i-},u\right)\right]=\lambda\int_0^t\mathbb{E}\left[f\left(Y^x_s,u\right)\right]\mathrm{d}s. \tag{3.4}$$

To this end, define, for $x,t\ge 0$ and $u>0$, $\eta_u(x,t)=\mathbb{E}[\sum_{i=1}^{N_t}f(Y^x_{T_i-},u)]$. With the help of the Markov property for (Y^x,N) and as well as the stationary independent increments of N, we have

$$\begin{aligned}\eta_u(x,t+s)-\eta_u(x,t)&=\mathbb{E}\left[\mathbb{E}\left[\sum_{i=N_t+1}^{N_{t+s}}f\left(Y^x_{T_i-},u\right)\middle|\mathcal{F}_t\right]\right]\\&=\mathbb{E}\left[\mathbb{E}\left[\sum_{i=1}^{\widetilde{N}_s}f\left(\widetilde{Y}^z_{\widetilde{T}_i-},u\right)\right]\Bigg|_{z=Y^x_t}\right]\\&=\mathbb{E}\left[\eta_u\left(Y^x_t,s\right)\right],\end{aligned}$$

where the process $\{(\widetilde{Y}^z_s,\widetilde{N}_s):s\ge 0\}$ is independent of $\mathcal{F}_t$ and equal in law to (Y^z,N) under $\mathbb{P}$. Note here that $\{\widetilde{T}_i:i\ge 1\}$ are the arrival times of the process $\widetilde{N}$. Next, note that, for all $x,s\ge 0$, we have that $s^{-1}\eta_u(x,s)$ is bounded by $s^{-1}C\mathbb{E}(N_s)=\lambda C$, where $C=\sup_{y\ge 0}f(y)<\infty$. Hence, with the help of the Dominated Convergence Theorem, our objective now is to compute the right-derivative of $\eta_u(x,t)$ by evaluating the limit

$$\lim_{s\downarrow 0}\frac{\eta_u(x,t+s)-\eta_u(x,t)}{s}=\mathbb{E}\left[\lim_{s\downarrow 0}\frac{1}{s}\eta_u\left(Y^x_t,s\right)\right]. \tag{3.5}$$

Note that, for all $x,s\ge 0$,

$$\frac{1}{s}\eta_u(x,s)=\frac{1}{s}\mathbb{E}\left[f\left(Y^x_{T_1-},u\right)\mathbf{1}_{(N_s=1)}\right]+\frac{1}{s}\mathbb{E}\left[\sum_{i\ge 1}^{N_s}f\left(Y^x_{T_i-},u\right)\mathbf{1}_{(N_s\ge 2)}\right]. \tag{3.6}$$

Moreover, recall that, for $v \le s$,

$$\begin{aligned}\mathbb{P}(T_1 \in \mathrm{d}v, N_s = 1) &= \mathbb{P}(T_1 \in \mathrm{d}v, T_2 > s)\\ &= \mathbb{P}(T_1 \in \mathrm{d}v, T_2 - T_1 > s - v)\\ &= \lambda \mathrm{e}^{-\lambda v}\mathrm{d}v \times \mathrm{e}^{-\lambda(s-v)}\\ &= \lambda \mathrm{e}^{-\lambda s}\mathrm{d}v.\end{aligned}$$

Hence, for the first term on the right-hand side of (3.6) we have

$$\lim_{s\downarrow 0}\frac{1}{s}\mathbb{E}\big[f\big(Y^x_{T_1-}, u\big)\mathbf{1}_{(N_s=1)}\big] = \lim_{s\downarrow 0}\lambda\frac{\mathrm{e}^{-\lambda s}}{s}\int_0^s f\big((x\vee cv) - cv, u\big)\mathrm{d}v = \lambda f(x, u).$$

For the second term on the right-hand side of (3.6), we also have

$$\begin{aligned}\lim_{s\downarrow 0}\frac{1}{s}\mathbb{E}\left[\sum_{i\ge 1}^{N_s} f\big(Y^x_{T_i-}, u\big)\mathbf{1}_{(N_s\ge 2)}\right] &\le C\lim_{s\downarrow 0}\frac{1}{s}\mathbb{E}[N_s\mathbf{1}_{(N_s\ge 2)}]\\ &= \lim_{s\downarrow 0}\frac{1}{s}\big[\lambda s\big(1 - \mathrm{e}^{-\lambda s}\big)\big] = 0.\end{aligned}$$

Returning to (3.6), it follows that, for all $x \ge 0$, $\lim_{s\downarrow 0} s^{-1}\eta_u(x, s) = \lambda f(x, u)$ and hence, from (3.5), we have that

$$\frac{\partial}{\partial t}\eta_u(x, t+) = \lambda\mathbb{E}\big[f\big(Y^x_t, u\big)\big].$$

A similar argument, looking at the difference $\eta_u(x, t) - \eta_u(x, t - s)$, for $x \ge 0$ and $t > s > 0$, also shows that the left derivative satisfies

$$\frac{\partial}{\partial t}\eta_u(x, t-) = \lambda\mathbb{E}\big[f\big(Y^x_t, u\big)\big].$$

It follows that $\eta_u(x, t)$ is differentiable in t on $(0, \infty)$ and hence, since $\eta_u(x, 0) = 0$,

$$\eta_u(x, t) = \lambda\int_0^t \mathbb{E}\big[f\big(Y^x_s, u\big)\big]\mathrm{d}s,$$

which establishes (3.4) and completes the proof. □

Remark 3.5 It is a straightforward exercise to deduce from Theorem 3.4 that the compensated process

$$\sum_{i=1}^{N_t} f\big(Y^x_{T_i-}, \xi_i\big) - \lambda\int_0^t\int_{(0,\infty)} \mathbb{E}\big(f\big(Y^x_s\big), u\big)F(\mathrm{d}u)\mathrm{d}s, \quad t \ge 0,$$

is a martingale.

3.3 Second Exponential Martingale

We are now ready to introduce our second exponential martingale, also known as the *Kella–Whitt* martingale. See Sect. 3.7 for historical remarks regarding its name.

Theorem 3.6 *For $\theta > 0$ and $x \geq 0$,*

$$M_t^x := \psi(\theta) \int_0^t \mathrm{e}^{-\theta Y_s^x} \mathrm{d}s + 1 - \mathrm{e}^{-\theta Y_t^x} - \theta(x \vee \overline{X}_t), \quad t \geq 0 \tag{3.7}$$

is a $\mathbb{P}$-martingale with respect to $\mathbb{F}$.

Proof Let us start by showing that

$$\mathbb{E}\big[|M_t^x|\big] < \infty,$$

for all $t, x \geq 0$. To this end, suppose that $\mathbf{e}_1$ is an independent and exponentially distributed random variable. We know from Corollary 2.6 that $\overline{X}_{\mathbf{e}_1}$ is exponentially distributed with parameter $\Phi(1)$. In particular, this implies that

$$\mathbb{E}[\overline{X}_{\mathbf{e}_1}] = \int_0^\infty \mathrm{e}^{-t} \mathbb{E}[\overline{X}_t] \mathrm{d}t < \infty.$$

Since $\overline{X}$ is an increasing process, it follows that $\mathbb{E}[\overline{X}_t] < \infty$, for all $t \geq 0$.

Using the triangle inequality for each of the terms in M_t^x, we may now estimate

$$\mathbb{E}\big[|M_t^x|\big] \leq \psi(\theta)t + 2 + \theta\mathbb{E}[x \vee \overline{X}_t] < \infty,$$

for $x, t \geq 0$.

Next, use the Markov property of Y^x to write, for $x, s, t \geq 0$,

$$x \vee \overline{X}_{t+s} = Y_{t+s}^x + X_{t+s} = \widetilde{Y}_s^z\big|_{z=Y_t^x} + \widetilde{X}_s + X_t = (z \vee \overline{\widetilde{X}}_s)|_{z=Y_t^x} + X_t,$$

where $\widetilde{X}_s = X_{t+s} - X_t$ and $\widetilde{Y}^z := (z \vee \overline{\widetilde{X}}) - \widetilde{X}$ is independent of $\{Y_u^x : u \leq t\}$ and equal in law to Y^z. Using this decomposition, it is straightforward to show that

$$\begin{aligned}\mathbb{E}\big[M_{t+s}^x|\mathcal{F}_t\big] &= \psi(\theta) \int_0^t \mathrm{e}^{-\theta Y_u^x} \mathrm{d}u + 1 - \theta X_t \\ &\quad + \mathbb{E}\left[\psi(\theta) \int_0^s \mathrm{e}^{-\theta Y_u^z} \mathrm{d}u - \mathrm{e}^{-\theta Y_s^z} - \theta(z \vee \overline{X}_s)\right]\Bigg|_{z=Y_t^x}.\end{aligned}$$

The proof is thus complete as soon as we show that, for all $z, s \geq 0$,

$$\mathbb{E}\left[\psi(\theta) \int_0^s \mathrm{e}^{-\theta Y_u^z} \mathrm{d}u - \mathrm{e}^{-\theta Y_s^z} - \theta(z \vee \overline{X}_s)\right] = -\mathrm{e}^{-\theta z} - \theta z.$$

In order to achieve this goal, we shall develop the left-hand side above using the so-called *chain rule* for right-continuous functions of bounded variation. We have that

$$\mathrm{e}^{-\theta Y_s^z} = \mathrm{e}^{-\theta z} - \theta \int_{(0,s]} \mathrm{e}^{-\theta Y_u^z} \mathrm{d}\big(Y_u^z\big)^c + \sum_{i=1}^{N_s} \big[\mathrm{e}^{-\theta Y_{T_i}^z} - \mathrm{e}^{-\theta Y_{T_i-}^z}\big], \tag{3.8}$$

for $z, s \geq 0$, where $(Y_u^z)^c$ is the continuous part of Y^z. Note that

$$\begin{aligned}
\int_{(0,s]} \mathrm{e}^{-\theta Y_u^z} \mathrm{d}\big(Y_u^z\big)^c &= \int_{(0,s]} \mathrm{e}^{-\theta Y_u^z} \mathrm{d}(z \vee \overline{X}_u) - \mathrm{c} \int_0^s \mathrm{e}^{-\theta Y_u^z} \mathrm{d}u \\
&= \int_{(0,s]} \mathbf{1}_{(Y_u^z = 0)} \mathrm{e}^{-\theta Y_u^z} \mathrm{d}(z \vee \overline{X}_u) - \mathrm{c} \int_0^s \mathrm{e}^{-\theta Y_u^z} \mathrm{d}u \\
&= (z \vee \overline{X}_s) - z - \mathrm{c} \int_0^s \mathrm{e}^{-\theta Y_u^z} \mathrm{d}u,
\end{aligned}$$

where in the second equality we have used the fact $Y_u^z = 0$ on the set of times that the process $z \vee \overline{X}_u$ increments. We may now take expectations in (3.8) to deduce that

$$\begin{aligned}
&\mathbb{E}\big[\mathrm{e}^{-\theta Y_s^z} + \theta(z \vee \overline{X}_s)\big] \\
&= \mathrm{e}^{-z\theta} + \theta z + \mathbb{E}\left[\mathrm{c}\theta \int_0^s \mathrm{e}^{-\theta Y_u^z} \mathrm{d}u + \sum_{i=1}^{N_s} \mathrm{e}^{-\theta Y_{T_i-}^z} \big(\mathrm{e}^{-\theta \xi_i} - 1\big)\right] \\
&= \mathrm{e}^{-z\theta} + \theta z + \mathrm{c}\theta \mathbb{E}\left[\int_0^s \mathrm{e}^{-\theta Y_u^z} \mathrm{d}u\right] \\
&\quad + \lambda \int_{(0,\infty)} \big(\mathrm{e}^{-\theta x} - 1\big) F(\mathrm{d}x) \mathbb{E}\left[\int_0^s \mathrm{e}^{-\theta Y_u^z} \mathrm{d}u\right] \\
&= \mathrm{e}^{-z\theta} + \theta z + \psi(\theta) \mathbb{E}\left[\int_0^s \mathrm{e}^{-\theta Y_u^z} \mathrm{d}u\right],
\end{aligned}$$

where we have applied Theorem 3.4 in the second equality. The proof is now complete. □

3.4 Duality

For our main application of the Kella–Whitt martingale, we need to address one additional property of the Cramér–Lundberg process, which follows as a consequence of the fact that it is also a Lévy process. This property concerns the issue of *duality*.

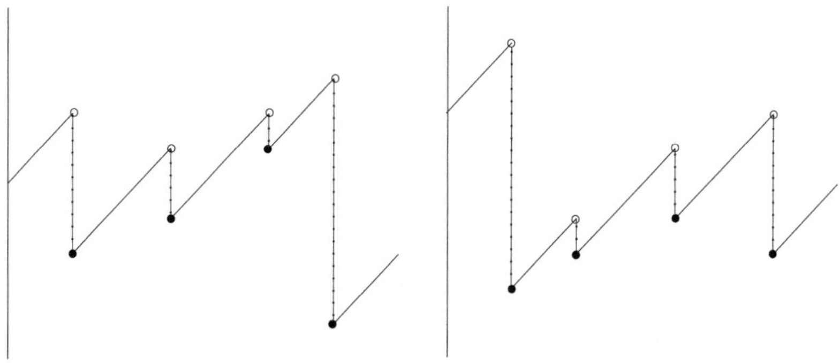

Fig. 3.1 A realisation of the trajectory of $\{X_s : 0 \leq s \leq t\}$ and of $\{X_{(t-s)-} - X_t : 0 \leq s \leq t\}$, respectively

Lemma 3.7 (Duality Lemma) *For each fixed $t > 0$, define the time-reversed process*

$$\{X_{(t-s)-} - X_t : 0 \leq s \leq t\}$$

and the dual process,

$$\{-X_s : 0 \leq s \leq t\}.$$

Then the two processes have the same law under $\mathbb{P}$.

Proof Define the process $R_s = X_t - X_{(t-s)-}$ for $0 \leq s \leq t$. Under $\mathbb{P}$, we have $R_0 = 0$ almost surely, as t is a jump time with probability zero. As can be seen from Fig. 3.1, the paths of R are obtained from those of X by a rotation through $180°$, with an adjustment of the continuity at the jump times, so that its paths are almost surely right-continuous with left limits. The stationary independent increments of X imply directly that the same is true of R. This puts R in the class of Lévy processes. Moreover, for each $0 \leq s \leq t$, the distribution of R_s is identical to that of X_s. It follows that

$$\mathbb{E}\left(e^{\lambda R_s}\right) = e^{\psi(\lambda)s},$$

for all $0 \leq s \leq t < \infty$ and $\lambda \geq 0$. Hence R has the same law as X. □

One interesting feature that follows as a consequence of the Duality Lemma is the relationship between the running supremum, the running infimum, the process reflected in its supremum and the process reflected in its infimum. The last four objects are, respectively,

$$\overline{X}_t = \sup_{0 \leq s \leq t} X_s, \qquad \underline{X}_t = \inf_{0 \leq s \leq t} X_s,$$

$$\{\overline{X}_t - X_t : t \geq 0\} \quad \text{and} \quad \{X_t - \underline{X}_t : t \geq 0\}.$$

Lemma 3.8 *For each fixed $t > 0$, the pairs $(\overline{X}_t, \overline{X}_t - X_t)$ and $(X_t - \underline{X}_t, -\underline{X}_t)$ have the same distribution under $\mathbb{P}$.*

Proof Define $R_s = X_t - X_{(t-s)-}$ for $0 \le s \le t$, as in the previous proof, and write $\underline{R}_t = \inf_{0\le s\le t} R_s$. Using right-continuity and left limits of paths we may deduce that

$$(\overline{X}_t, \overline{X}_t - X_t) = (R_t - \underline{R}_t, -\underline{R}_t)$$

almost surely. Now appealing to the Duality Lemma we have that $\{R_s : 0 \le s \le t\}$ is equal in law to $\{X_s : 0 \le s \le t\}$ under $\mathbb{P}$ and the result follows. □

3.5 Distribution of the Minimum

We are now able to deliver the promised result concerning the law of the minimum.

Theorem 3.9 *Let $\underline{X}_t = \inf_{0\le u\le t} X_u$ and suppose that $\mathbf{e}_q$ is an exponentially distributed random variable with parameter $q > 0$, which is independent of the process X. Then, for $\theta > 0$,*

$$\mathbb{E}\big(\mathrm{e}^{\theta \underline{X}_{\mathbf{e}_q}}\big) = \frac{q(\theta - \Phi(q))}{\Phi(q)(\psi(\theta) - q)}, \tag{3.9}$$

where the right-hand side is understood in the asymptotic sense when $\theta = \Phi(q)$, i.e. $q/\Phi(q)\psi'(\Phi(q))$.

Proof Let us first consider the case that $\theta, q > 0$ and $\theta \ne \Phi(q)$. Let $Y_t = Y_t^0 = \overline{X}_t - X_t$. By an application of Fubini's Theorem together with Lemma 3.8,

$$\mathbb{E}\left[\int_0^{\mathbf{e}_q} \mathrm{e}^{-\theta Y_s}\mathrm{d}s\right] = \int_0^\infty \mathrm{e}^{-qs}\mathbb{E}\big(\mathrm{e}^{-\theta Y_s}\big)\mathrm{d}s = \frac{1}{q}\mathbb{E}\big(\mathrm{e}^{-\theta Y_{\mathbf{e}_q}}\big) = \frac{1}{q}\mathbb{E}\big(\mathrm{e}^{\theta \underline{X}_{\mathbf{e}_q}}\big).$$

From Theorem 3.6, we have that $\mathbb{E}(M^0_{\mathbf{e}_q}) = \mathbb{E}(M^0_0) = 0$, and hence we obtain

$$\frac{\psi(\theta) - q}{q}\mathbb{E}\big(\mathrm{e}^{\theta \underline{X}_{\mathbf{e}_q}}\big) = \theta\mathbb{E}(\overline{X}_{\mathbf{e}_q}) - 1.$$

Recall from Corollary 2.6 that $\overline{X}_{\mathbf{e}_q}$ is exponentially distributed with parameter $\Phi(q)$ and hence $\mathbb{E}(\overline{X}_{\mathbf{e}_q}) = 1/\Phi(q)$. It follows that

$$\frac{\psi(\theta) - q}{q}\mathbb{E}\big(\mathrm{e}^{\theta \underline{X}_{\mathbf{e}_q}}\big) = \frac{\theta - \Phi(q)}{\Phi(q)}. \tag{3.10}$$

For the case that $q > 0$ and $\theta = \Phi(q)$, the result follows from the case that $\theta \ne \Phi(q)$ by taking limits as $\theta \to \Phi(q)$. □

3.6 The Long-Term Behaviour

Let us conclude this chapter by returning to earlier remarks made in Sect. 1.2 regarding the long-term behaviour of the Cramér–Lundberg process. Recall that $\psi'(0+) = c - \lambda\mu$, where c is the premium rate, λ is the rate of claim arrivals and μ is their common mean. It is clear from (1.4) that, when $\psi'(0+) > 0$, we have $\lim_{t\to\infty} X_t = \infty$ and when $\psi'(0+) < 0$, $\lim_{t\to\infty} X_t = -\infty$. For the remaining case, when $\psi'(0+) = 0$, the Strong Law of Large Numbers is not as informative. We can, however, use our previous results on the law of the maximum and minimum of X to determine the long-term behaviour of X. Specifically, the lemma below shows that, when $\psi'(0+) = 0$, the process X *oscillates* in the sense that $\limsup_{t\to\infty} X_t = -\liminf_{t\to\infty} X_t = \infty$.

Lemma 3.10 *We have that*

(i) $\overline{X}_\infty$ *and* $-\underline{X}_\infty$ *are either infinite almost surely or finite almost surely,*
(ii) $\overline{X}_\infty = \infty$ *if and only if* $\psi'(0^+) \geq 0$,
(iii) $\underline{X}_\infty = -\infty$ *if and only if* $\psi'(0^+) \leq 0$.

Proof On account of the strict convexity ψ, we have that $\Phi(0) > 0$ if and only if $\psi'(0^+) < 0$. Hence,

$$\lim_{q\downarrow 0} \frac{q}{\Phi(q)} = \begin{cases} 0 & \text{if } \psi'(0^+) \leq 0, \\ \psi'(0^+) & \text{if } \psi'(0^+) > 0. \end{cases}$$

In the case that $\psi'(0+) \geq 0$ above, i.e. the case $\Phi(0) = 0$, to compute the limit, we make use of the fact that $q/\Phi(q)$ can otherwise be written $\psi(\Phi(q))/\Phi(q)$. By taking the limit as q tends to zero in the identity (3.9), we now have that

$$\mathbb{E}\big(e^{\theta \underline{X}_\infty}\big) = \begin{cases} 0 & \text{if } \psi'(0^+) \leq 0, \\ \psi'(0^+)\theta/\psi(\theta) & \text{if } \psi'(0^+) > 0. \end{cases} \tag{3.11}$$

In the first of the two cases above, it is clear that $\mathbb{P}(-\underline{X}_\infty = \infty) = 1$. In the second case, taking limits as $\theta \to \infty$, one sees that $\mathbb{P}(-\underline{X}_\infty = \infty) = 0$.

Next, recall from Corollary 2.6 that $\overline{X}_\infty$ is exponentially distributed with parameter $\Phi(0)$. In particular, $\overline{X}_\infty$ is almost surely infinite when $\psi'(0+) \geq 0$ and almost surely finite when $\psi'(0+) < 0$.

Putting this information together, the statements (i)–(iii) are easily recovered. □

3.7 Comments

The fact that the process Y^x is a Markov process for each $x \geq 0$ is well known from queueing theory, where the process Y^x can be seen as the workload in an $M/G/1$ queue (when the initial workload at time zero is x). The proof given here is taken from Bingham (1975). Theorem 3.4 is an example of the so-called compensation formula which can be stated for general Poisson integrals. See for example Chapter XII.1 of Revuz and Yor (2004) or Chapter 0.5 of Bertoin (1996). The Kella–Whitt martingale was first introduced in Kella and Whitt (1992) in the setting of a general Lévy process. It is closely related to the so-called Kennedy and Azéma–Yor martingales, both of which have previously been studied in the setting of Brownian motion. See Azéma and Yor (1979) and Kennedy (1976). A general version of the chain rule can be found in Proposition (4.6) of Chapter 0 in Revuz and Yor (2004). The Duality Lemma is also well known for (and in fact originates from) the theory of random walks and is justified using an almost identical proof. See for example Chapter XII of Feller (1971) for random walks and Chapter II.1 of Bertoin (1996) for Lévy processes.

Chapter 4
Scale Functions and Ruin Probabilities

The two main results from the previous chapters concerning the law of the maximum and minimum of the Cramér–Lundberg process can now be put to use in order to establish our first results concerning the classical ruin problem. We shall introduce the so-called *scale functions*, which will prove to be indispensable, both in this chapter and later, when describing various distributional features of the ruin problem.

4.1 Scale Functions and the Probability of Ruin

For a given Cramér–Lundberg process X with Laplace exponent ψ, we want to define a family of functions, indexed by $q \geq 0$, which we shall denote by $W^{(q)} : \mathbb{R} \to [0, \infty)$. For all $q \geq 0$ we shall set $W^{(q)}(x) = 0$ for $x < 0$. The next theorem serves as a definition for $W^{(q)}$ on $[0, \infty)$.

Theorem 4.1 *For all $q \geq 0$ there exists a function $W^{(q)}$ on $[0, \infty)$ defined to be the unique non-decreasing, right-continuous function whose Laplace transform is given by*

$$\int_0^\infty \mathrm{e}^{-\beta x} W^{(q)}(x)\mathrm{d}x = \frac{1}{\psi(\beta) - q}, \quad \beta > \Phi(q). \tag{4.1}$$

For convenience, we shall always write W in place of $W^{(0)}$. Typically, we shall refer to the functions $W^{(q)}$ as *q-scale functions*, but we shall also refer to W as just the *scale function*.

Proof of Theorem 4.1 First assume that $\psi'(0+) > 0$. With a pre-emptive choice of notation, we shall define the function

$$W(x) = \frac{1}{\psi'(0+)} \mathbb{P}_x(\underline{X}_\infty \geq 0), \quad x \in \mathbb{R}. \tag{4.2}$$

A.E. Kyprianou, *Gerber–Shiu Risk Theory*, EAA Series,
DOI 10.1007/978-3-319-02303-8_4,

Clearly $W(x) = 0$ for $x < 0$ and it is non-decreasing and right-continuous since it is also proportional to the distribution function $\mathbb{P}(-\underline{X}_\infty \leq x)$. Integration by parts shows that, on the one hand,

$$\begin{aligned}\int_0^\infty \mathrm{e}^{-\beta x} W(x)\mathrm{d}x &= \frac{1}{\psi'(0+)} \int_0^\infty \mathrm{e}^{-\beta x}\mathbb{P}(-\underline{X}_\infty \leq x)\mathrm{d}x \\ &= \frac{1}{\psi'(0+)\beta} \int_{[0,\infty)} \mathrm{e}^{-\beta x}\mathbb{P}(-\underline{X}_\infty \in \mathrm{d}x) \\ &= \frac{1}{\psi'(0+)\beta}\mathbb{E}\big(\mathrm{e}^{\beta \underline{X}_\infty}\big). \qquad (4.3)\end{aligned}$$

On the other hand, recalling (3.11), we also have that

$$\mathbb{E}\big(\mathrm{e}^{\beta \underline{X}_\infty}\big) = \frac{\psi'(0+)\beta}{\psi(\beta)}, \quad \beta \geq 0.$$

When combined with (4.3), this gives us (4.1) for the case $q = 0$ and $\psi'(0+) > 0$.

Next, we deal with the case that $q > 0$ or that $q = 0$ *and* $\psi'(0+) < 0$. To this end, again making use of a pre-emptive choice of notation, let us define the non-decreasing and right-continuous function

$$W^{(q)}(x) = \mathrm{e}^{\Phi(q)x} W_{\Phi(q)}(x), \quad x \geq 0, \qquad (4.4)$$

where $W_{\Phi(q)}$ plays the role of W, but now for the process $(X, \mathbb{P}^{\Phi(q)})$. Note in particular that, by Theorem 2.3, $(X, \mathbb{P}^{\Phi(q)})$ has Laplace exponent

$$\psi_{\Phi(q)}(\theta) = \psi\big(\theta + \Phi(q)\big) - q, \quad \theta \geq 0. \qquad (4.5)$$

Hence $\psi'_{\Phi(q)}(0+) = \psi'(\Phi(q)) > 0$, which ensures that $W_{\Phi(q)}$ is well defined by the previous part of the proof. Taking Laplace transforms, we have, for $\beta > \Phi(q)$,

$$\begin{aligned}\int_0^\infty \mathrm{e}^{-\beta x} W^{(q)}(x)\mathrm{d}x &= \int_0^\infty \mathrm{e}^{-(\beta - \Phi(q))x} W_{\Phi(q)}(x)\mathrm{d}x \\ &= \frac{1}{\psi_{\Phi(q)}(\beta - \Phi(q))} \\ &= \frac{1}{\psi(\beta) - q},\end{aligned}$$

thus completing the proof for the case that $q > 0$ or that $q = 0$ and $\psi'(0+) < 0$.

Finally, we deal with the case that $q = 0$ and $\psi'(0+) = 0$. Since $W_{\Phi(q)}(x)$ is an increasing function, we may also treat it as a distribution function of a measure which, with as an abuse of notation, we also call $W_{\Phi(q)}$. Integrating by parts thus gives us, for $\beta > 0$,

$$\int_{[0,\infty)} \mathrm{e}^{-\beta x} W_{\Phi(q)}(\mathrm{d}x) = \frac{\beta}{\psi_{\Phi(q)}(\beta)}. \qquad (4.6)$$

Note that the assumption $\psi'(0+) = 0$ implies that $\Phi(0) = 0$ and hence, for $\theta \geq 0$,

$$\lim_{q \downarrow 0} \psi_{\Phi(q)}(\theta) = \lim_{q \downarrow 0} \big[\psi\big(\theta + \Phi(q)\big) - q\big] = \psi(\theta).$$

One may appeal to (4.6) and the Extended Continuity Theorem for Laplace transforms, see for example Theorem XIII.1.2a of Feller (1971), to deduce that, since

$$\lim_{q \downarrow 0} \int_{[0,\infty)} \mathrm{e}^{-\beta x} W_{\Phi(q)}(\mathrm{d}x) = \frac{\beta}{\psi(\beta)},$$

there exists a measure W^* such that $W^*(x) := W^*[0,x] = \lim_{q \downarrow 0} W_{\Phi(q)}(x)$ and

$$\int_{[0,\infty)} \mathrm{e}^{-\beta x} W^*(\mathrm{d}x) = \frac{\beta}{\psi(\beta)}.$$

Integration by parts shows that W^* satisfies

$$\int_0^\infty \mathrm{e}^{-\beta x} W^*(x)\mathrm{d}x = \frac{1}{\psi(\beta)},$$

for $\beta > 0$, as required. It is clear from its definition that $W := W^*$ is non-decreasing, right-continuous and satisfies (4.1). □

With the definition of scale functions in hand, we can return to the problem of ruin. The following corollary follows as a simple consequence of Laplace inversion of the identity in Theorem 3.9, taking account of (4.1).

Corollary 4.2 *For* $x, q > 0$,

$$\mathbb{P}(-\underline{X}_{\mathbf{e}_q} \in \mathrm{d}x) = \frac{q}{\Phi(q)} W^{(q)}(\mathrm{d}x) - q W^{(q)}(x)\mathrm{d}x. \tag{4.7}$$

In the above formula, thanks to (4.4), the function $W^{(q)}$ is increasing and hence the measure $W^{(q)}(\mathrm{d}x)$, $x \geq 0$, makes sense (albeit being another abuse of notation). Formula (4.7) can also be stated when $q = 0$, providing $\psi'(0+) > 0$. In that case, as we have seen before, the term $q/\Phi(q)$ should be understood in the limiting sense as equal to $\psi'(0+)$.

We complete this section with our main result about ruin probabilities, using scale functions. To this end, let us define the functions

$$Z^{(q)}(x) = 1 + q \int_0^x W^{(q)}(y)\mathrm{d}y, \quad x \in \mathbb{R},$$

for $q \geq 0$. Moreover, recall that

$$\tau_0^- := \inf\{t > 0 : X_t < 0\}$$

and that $\mathrm{e}^{-\infty} := 0$.

Theorem 4.3 (Time to ruin) *For any $x \in \mathbb{R}$ and $q > 0$,*

$$\mathbb{E}_x\left(e^{-q\tau_0^-}\right) = Z^{(q)}(x) - \frac{q}{\Phi(q)} W^{(q)}(x). \tag{4.8}$$

By taking limits above as $q \to 0$, we also have

$$\mathbb{P}_x\left(\tau_0^- < \infty\right) = \begin{cases} 1 - \psi'(0+)W(x) & \text{if } \psi'(0+) \geq 0 \\ 1 & \text{if } \psi'(0+) < 0. \end{cases} \tag{4.9}$$

Proof Appealing to (4.7), we have, for $x \geq 0$,

$$\begin{aligned} \mathbb{E}_x\left(e^{-q\tau_0^-}\right) &= \mathbb{P}_x\left(\mathbf{e}_q > \tau_0^-\right) \\ &= \mathbb{P}_x(\underline{X}_{\mathbf{e}_q} < 0) \\ &= \mathbb{P}(-\underline{X}_{\mathbf{e}_q} > x) \\ &= 1 - \mathbb{P}(-\underline{X}_{\mathbf{e}_q} \leq x) \\ &= 1 + q\int_0^x W^{(q)}(y)\mathrm{d}y - \frac{q}{\Phi(q)} W^{(q)}(x) \\ &= Z^{(q)}(x) - \frac{q}{\Phi(q)} W^{(q)}(x). \end{aligned} \tag{4.10}$$

Note that, since $Z^{(q)}(x) = 1$ and $W^{(q)}(x) = 0$ for all $x \in (-\infty, 0)$, the statement is valid for all $x \in \mathbb{R}$. The proof is now complete for the case that $q > 0$.

In order to deal with the case $q = 0$, recall our previous trick in writing

$$\lim_{q \downarrow 0} q/\Phi(q) = \lim_{q \downarrow 0} \psi\left(\Phi(q)\right)/\Phi(q).$$

If $\psi'(0+) \geq 0$, i.e. the process drifts to infinity or oscillates, then $\Phi(0) = 0$ and the limit is equal to $\psi'(0+)$. Otherwise, when $\Phi(0) > 0$, the aforementioned limit is zero. The proof is thus completed by taking the limit in q in (4.8). □

The last part of the above theorem can also be recovered directly from the definition of W in the case that $\psi'(0+) > 0$, see (4.2). Moreover, given the discussion in Sect. 3.6, the probability of ruin when $\psi'(0+) \leq 0$ is obviously 1.

4.2 Connection with the Pollaczek–Khintchine Formula

In Theorem 1.3, we gave the classical Pollaczek–Khintchine formula for the probability of ruin in the case that $\psi'(0+) > 0$. Compared with (4.9), it is not immediately obvious how these two formulae relate to one another. Let us therefore spend a little time to make the connection between the two, first with an analytical explanation and then with a probabilistic one.

Analytical Explanation Let us start by noting that, just as in formula (4.6), we can integrate by parts the Laplace transform of W to show that

$$\int_{[0,\infty)} \mathrm{e}^{-\beta x} W(\mathrm{d}x) = \frac{\beta}{\psi(\beta)}, \quad \beta > 0.$$

Alternatively, this also follows from the definition (4.2) and the expression for the Laplace transform of $-\underline{X}_\infty$, given in (3.11). Next, note that $\psi'(0+) = \mathrm{c} - \lambda\mu > 0$ implies that $\mu < \infty$ and that

$$\rho := \lambda\mu/\mathrm{c} < 1.$$

This inequality also implies that

$$\frac{\lambda\mu}{\mathrm{c}} \int_0^\infty \mathrm{e}^{-\beta x} \frac{1}{\mu} \overline{F}(x)\mathrm{d}x < 1,$$

where we recall that $\overline{F}(x) = 1 - F(x)$. Hence, recalling the representation of ψ given in (2.4), we can write, for $\beta > 0$,

$$\begin{aligned} \frac{\beta}{\psi(\beta)} &= \frac{1}{\mathrm{c}} \frac{1}{1 - \frac{\lambda\mu}{\mathrm{c}} \int_0^\infty \mathrm{e}^{-\beta x} \frac{1}{\mu} \overline{F}(x)\mathrm{d}x} \\ &= \frac{1}{\mathrm{c}} \sum_{k=0}^\infty \rho^k \left(\int_0^\infty \mathrm{e}^{-\beta x} \frac{1}{\mu} \overline{F}(x)\mathrm{d}x \right)^k. \end{aligned} \quad (4.11)$$

Next note that

$$\eta(\mathrm{d}x) := \frac{1}{\mu} \overline{F}(x)\mathrm{d}x, \quad x \geq 0,$$

is a probability measure. For each $k \geq 0$, recall that $\eta^{*k}(\mathrm{d}x)$, $x \geq 0$, its k-fold convolution, where we understand $\eta^{*0}(\mathrm{d}x) := \delta_0(\mathrm{d}x)$, $x \geq 0$, the Dirac delta measure which places an atom of unit mass at zero. Since, for $\beta > 0$ and $k \geq 0$,

$$\int_{[0,\infty)} \mathrm{e}^{-\beta x} \eta^{*k}(\mathrm{d}x) = \left(\int_0^\infty \mathrm{e}^{-\beta x} \frac{1}{\mu} \overline{F}(x)\mathrm{d}x \right)^k,$$

we may apply Laplace inversion to the right-hand side of (4.11) and conclude that, for $x \geq 0$,

$$W(\mathrm{d}x) = \frac{1}{\mathrm{c}} \sum_{k=0}^\infty \rho^k \eta^{*k}(\mathrm{d}x).$$

Said another way, we have, for $x \geq 0$,

$$W(x) = \frac{1}{\mathrm{c}} \sum_{k=0}^\infty \rho^k \eta^{*k}(x). \quad (4.12)$$

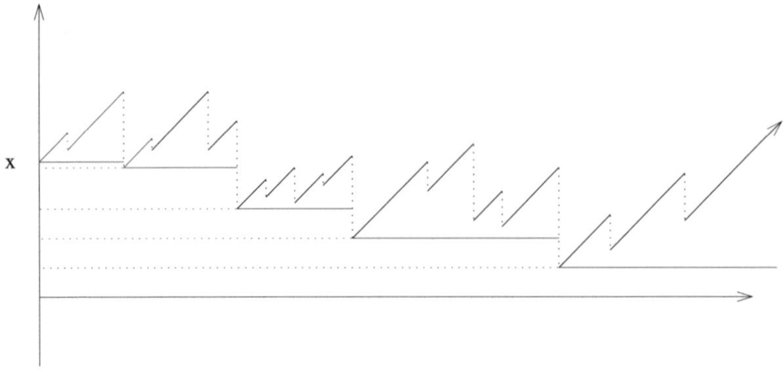

Fig. 4.1 A path of the Cramér–Lundberg process which drifts to ∞ before passing below 0. The *horizontal line* segments mark the successive minima. The *vertical distances* between subsequent *horizontal line* segments represent the quantities Δ_n

Returning to the formula in (4.9), when $\psi'(0+) = c - \lambda\mu > 0$, we now see that

$$1 - \mathbb{P}_x\big(\tau_0^- < \infty\big) = (1-\rho)\sum_{k=0}^{\infty}\rho^k \eta^{*k}(x), \tag{4.13}$$

as stated in Theorem 1.3.

Probabilistic Explanation The Pollaczek–Khintchine formula can also be recovered by looking at the successive minima of the process X. To this end, let us set $\Theta_0 = 0$ and sequentially define, for all $k \geq 1$ such that $\Theta_{k-1} < \infty$,

$$\Theta_k = \inf\{t > \Theta_{k-1} : X_t < X_{\Theta_{k-1}}\},$$

with the usual understanding that $\inf\emptyset = \infty$. As long as they are finite, the times Θ_k are the times of successive new minima.

The strong Markov property implies that, for each $k \geq 1$ such that $\{\Theta_{k-1} < \infty\}$, the pair $(\Theta_k - \Theta_{k-1}, X_{\Theta_k} - X_{\Theta_{k-1}})$ is independent of $\mathcal{F}_{\Theta_{k-1}}$ and equal in law to the pair $(\tau_0^-, X_{\tau_0^-})$, where we understand $X_{\Theta_k} := \infty$ when $\Theta_k = \infty$ and, similarly, $X_{\tau_0^-} := \infty$ when $\tau_0^- = \infty$. For $k \geq 1$, define on $\{\Theta_{k-1} < \infty\}$

$$\Delta_k = -(X_{\Theta_k} - X_{\Theta_{k-1}}).$$

The event $\{\tau_0^- = \infty\}$ under $\mathbb{P}_x$, $x \geq 0$, corresponds to the event that

$$\left\{\sum_{n=1}^{\nu-1}\Delta_n \leq x\right\},$$

where $\nu = \min\{k \geq 1 : \Theta_k = \infty\}$. See Fig. 4.1. By the strong Markov property, the index ν is the time of first success in an independent sequence of Bernoulli trials with probability of "success" $1-\widehat{\rho} := \mathbb{P}(\Theta_1 = \infty) = \mathbb{P}(\tau_0^- = \infty)$. In other words, ν is geometrically distributed. Moreover, ν is independent of the outcome of each of trials preceding the ν-th trial. Each of these trails "fails", delivering a random value which is distributed according to the measure $\widehat{\eta}(\mathrm{d}x) := \mathbb{P}(\Delta_1 \in \mathrm{d}x) = \mathbb{P}(-X_{\tau_0^-} \in \mathrm{d}x | \tau_0^- < \infty)$, $x > 0$.

In conclusion, we see that

$$1 - \mathbb{P}_x\big(\tau_0^- < \infty\big) = (1-\widehat{\rho}) \sum_{k\geq 0} \widehat{\rho}^k \widehat{\eta}^{*k}(x), \quad x \geq 0. \tag{4.14}$$

Comparing the formulae (4.13) and (4.14) when $x = 0$, we see that $\mathbb{P}(\tau_0^- < \infty) = \rho = \widehat{\rho}$, and hence it follows that $\eta = \widehat{\eta}$.

The following corollary falls straight out of the above comparison.

Corollary 4.4 *If* $\psi'(0+) > 0$, *then*

$$\mathbb{P}\big(\tau_0^- < \infty\big) = \frac{\lambda\mu}{\mathrm{c}} \quad \textit{and} \quad \mathbb{P}\big(-X_{\tau_0^-} \leq x | \tau_0^- < \infty\big) = \frac{1}{\mu}\int_0^x \overline{F}(y)\mathrm{d}y, \quad x \geq 0.$$

4.3 Gambler's Ruin

A slightly more elaborate version of the ruin problem is to consider the event that a certain surplus level, say $a \geq 0$, can be achieved before ruin. This is also known as the *gambler's ruin problem.*

Theorem 4.5 *For all* $q \geq 0$, $a > 0$ *and* $x \leq a$,

$$\mathbb{E}_x\big(\mathrm{e}^{-q\tau_a^+}\mathbf{1}_{(\tau_a^+ < \tau_0^-)}\big) = \frac{W^{(q)}(x)}{W^{(q)}(a)}. \tag{4.15}$$

Proof First, we deal with the case that $q = 0$ and $\psi'(0+) > 0$ as in the previous proof. Since we have identified $W(x) = \mathbb{P}_x(\underline{X}_\infty \geq 0)/\psi'(0+)$, a simple argument, using the law of total probability and the strong Markov property, now yields, for $x \in [0,a]$,

$$\begin{aligned}
&\mathbb{P}_x(\underline{X}_\infty \geq 0) \\
&\quad = \mathbb{E}_x\big(\mathbb{P}_x(\underline{X}_\infty \geq 0 | \mathcal{F}_{\tau_a^+})\big) \\
&\quad = \mathbb{E}_x\big(\mathbf{1}_{(\tau_a^+ < \tau_0^-)}\mathbb{P}_a(\underline{X}_\infty \geq 0)\big) + \mathbb{E}_x\big(\mathbf{1}_{(\tau_a^+ > \tau_0^-)}\mathbb{P}_{X_{\tau_0^-}}(\underline{X}_\infty \geq 0)\big).
\end{aligned} \tag{4.16}$$

The first term on the right-hand side of (4.16) is equal to

$$\mathbb{P}_a(\underline{X}_\infty \geq 0)\mathbb{P}_x\big(\tau_a^+ < \tau_0^-\big).$$

The second term on the right-hand side of (4.16) turns out to be equal to zero. To see why, note that $X_{\tau_0^-} < 0$ and the claim follows by virtue of the fact that $\mathbb{P}_x(\underline{X}_\infty \geq 0) = 0$ for all $x < 0$. We may now deduce that

$$\mathbb{P}_x\big(\tau_a^+ < \tau_0^-\big) = \frac{W(x)}{W(a)}, \quad x \in [0, a], \tag{4.17}$$

and clearly the same equality holds even when $x < 0$ as both left- and right-hand side are identically equal to zero.

Next, we deal with the case $q > 0$. Making use of the Esscher transform and recalling that $X_{\tau_a^+} = a$ on $\{\tau_a^+ < \infty\}$, we have that

$$\begin{aligned}
\mathbb{E}_x\big(\mathrm{e}^{-q\tau_a^+}\mathbf{1}_{(\tau_a^+ < \tau_0^-)}\big) &= \mathrm{e}^{-\Phi(q)(a-x)}\mathbb{E}_x\big(\mathrm{e}^{\Phi(q)(X_{\tau_a^+}-x)-q\tau_a^+}\mathbf{1}_{(\tau_a^+ < \tau_0^-)}\big) \\
&= \mathrm{e}^{-\Phi(q)(a-x)}\mathbb{P}_x^{\Phi(q)}\big(\tau_a^+ < \tau_0^-\big) \\
&= \mathrm{e}^{-\Phi(q)(a-x)}\frac{W_{\Phi(q)}(x)}{W_{\Phi(q)}(a)} \\
&= \frac{W^{(q)}(x)}{W^{(q)}(a)}.
\end{aligned}$$

Finally, to deal with the case that $q = 0$ and $\psi'(0+) \leq 0$, one needs only to take limits as $q \downarrow 0$ in the above identity, making use of monotone convergence on the left-hand side and continuity in q on the right-hand side, using the Continuity Theorem for Laplace transforms. □

We can also consider the converse event that ruin occurs prior to achieving a desired surplus of $a \geq 0$.

Theorem 4.6 *For any $x \leq a$ and $q \geq 0$,*

$$\mathbb{E}_x\big(\mathrm{e}^{-q\tau_0^-}\mathbf{1}_{(\tau_0^- < \tau_a^+)}\big) = Z^{(q)}(x) - Z^{(q)}(a)\frac{W^{(q)}(x)}{W^{(q)}(a)}. \tag{4.18}$$

Proof Fix $q > 0$. We have for $x \geq 0$,

$$\mathbb{E}_x\big(\mathrm{e}^{-q\tau_0^-}\mathbf{1}_{(\tau_0^- < \tau_a^+)}\big) = \mathbb{E}_x\big(\mathrm{e}^{-q\tau_0^-}\big) - \mathbb{E}_x\big(\mathrm{e}^{-q\tau_0^-}\mathbf{1}_{(\tau_a^+ < \tau_0^-)}\big).$$

Applying the strong Markov property at τ_a^+ and using the fact that $X_{\tau_a^+} = a$ on $\{\tau_a^+ < \infty\}$, we also have that

$$\mathbb{E}_x\big(\mathrm{e}^{-q\tau_0^-}\mathbf{1}_{(\tau_a^+ < \tau_0^-)}\big) = \mathbb{E}_x\big(\mathrm{e}^{-q\tau_a^+}\mathbf{1}_{(\tau_a^+ < \tau_0^-)}\big)\mathbb{E}_a\big(\mathrm{e}^{-q\tau_0^-}\big).$$

Appealing to (4.8) and (4.15), we now have that

$$\begin{aligned}\mathbb{E}_x\left(\mathrm{e}^{-q\tau_0^-}\mathbf{1}_{(\tau_0^-<\tau_a^+)}\right) &= Z^{(q)}(x) - \frac{q}{\Phi(q)}W^{(q)}(x) \\ &\quad - \frac{W^{(q)}(x)}{W^{(q)}(a)}\left(Z^{(q)}(a) - \frac{q}{\Phi(q)}W^{(q)}(a)\right),\end{aligned}$$

and the required result follows in the case that $q > 0$. The case that $q = 0$ is again dealt with by taking limits as $q \downarrow 0$. □

4.4 Comments

The name "scale function" for W was first used in Bertoin (1992) to reflect the analogous role it plays in (4.15) to scale functions for diffusions. The gambler's ruin problem (also known as the two-sided exit problem) has a long history, starting with the early work in Zolotarev (1964) and Takács (1966), followed by Rogers (1990), all of whom dealt with (4.15) in the case $q = 0$. The case that $q > 0$ was dealt with in Gerber (1972), Korolyuk (1975a), and later by Bertoin (1997). A recent summary of the theory of scale functions and its applications can be found in Cohen et al. (2013).

Chapter 5
The Gerber–Shiu Measure

Having introduced scale functions, we are now ready to look at the Gerber–Shiu measure in detail. In this chapter, we shall develop an idea from the previous chapter, involving Bernoulli trials of excursions from the minimum, to provide an identity for the expected occupation measure until ruin of the Cramér–Lundberg process. This identity will then play a key role in identifying an expression for the Gerber–Shiu measure. In fact, the analysis we give will work equally well in the context of the gambler's ruin problem.

5.1 Decomposing Paths at the Minimum

The main objective in this section is to prove the following decoupling for the path of the Cramér–Lundberg process when sampled at an independent and exponentially distributed random time, $\mathbf{e}_q$, with rate $q > 0$.

Theorem 5.1 *For all $q > 0$, the random variables $X_{\mathbf{e}_q} - \underline{X}_{\mathbf{e}_q}$ and $\underline{X}_{\mathbf{e}_q}$ are independent.*

Proof The proof makes use of the path decomposition we previously employed for the probabilistic explanation of the Pollaczek–Khintchine formula. Recall that, in that setting, we sequentially defined $\Theta_0 = 0$ and, for all $k \geq 1$ such that $\Theta_{k-1} < \infty$,

$$\Theta_k = \inf\{t > \Theta_{k-1} : X_t < X_{\Theta_{k-1}}\}.$$

Moreover, on the event $\{\Theta_{k-1} < \infty\}$ we defined $\Delta_k = -(X_{\Theta_k} - X_{\Theta_{k-1}})$. Formally speaking, the times $\Theta_1, \Theta_2, \ldots$ describe the right endpoints of *excursions from the minimum*. The latter may be thought of as the sequence of segments of the trajectory of X given by $\{X_{\Theta_{k-1}+t} : t \in (0, \Theta_k]\}$ for all $k \geq 1$ such that $\Theta_{k-1} < \infty$.

Although the discussion in the context of the Pollaczek–Khintchine formula focused exclusively on the case that $\psi'(0+) > 0$, the times Θ_k are still well defined

A.E. Kyprianou, *Gerber–Shiu Risk Theory*, EAA Series,
DOI 10.1007/978-3-319-02303-8_5,

when $\psi'(0+) \leq 0$. In fact, in this regime, since $\underline{X}_\infty = -\infty$ almost surely, it follows that $\Theta_k < \infty$ almost surely for all $k \geq 1$.

Now suppose that $\{\mathbf{e}_q^{(k)} : k \geq 1\}$ is a sequence of independent and identically distributed exponential random variables with parameter $q > 0$. Let

$$\ell := \min\{k \geq 1 : \Theta_k - \Theta_{k-1} > \mathbf{e}_q^{(k)}\}$$

be the index of the first excursion from the minimum which, in duration, exceeds the correspondingly indexed exponential random variable in the sequence $\{\mathbf{e}_q^{(k)} : k \geq 1\}$. By the lack of memory property, we can now identify the pair $(X_{\mathbf{e}_q} - \underline{X}_{\mathbf{e}_q}, -\underline{X}_{\mathbf{e}_q})$ as equal in law to the pair

$$\left(X_{\Theta_{\ell-1}+\mathbf{e}_q^{(\ell)}} - X_{\Theta_{\ell-1}}, \sum_{j=1}^{\ell-1} \Delta_j\right). \tag{5.1}$$

Appealing again to the concept of Bernoulli trials, it is clear that both the random variable ℓ and the ℓ-th excursion from the minimum will be independent of the preceding $\ell - 1$ excursions from the minimum. In particular, this implies that the pair in (5.1) is independent, and hence so is the pair $(X_{\mathbf{e}_q} - \underline{X}_{\mathbf{e}_q}, -\underline{X}_{\mathbf{e}_q})$. □

Note that the above proof allows us to say a little more about $(X_{\mathbf{e}_q} - \underline{X}_{\mathbf{e}_q}, -\underline{X}_{\mathbf{e}_q})$, thanks to the representation (5.1). Indeed, $X_{\mathbf{e}_q} - \underline{X}_{\mathbf{e}_q}$ is equal in law to $X_{\mathbf{e}_q}$ conditional on $\{\mathbf{e}_q < \tau_0^-\}$. Moreover, all of the Δ_j in the sum in (5.1) are i.i.d. and equal in distribution to $-X_{\tau_0^-}$ conditional on $\{\tau_0^- < \mathbf{e}_q\}$, and ℓ is independent and geometrically distributed with parameter $\mathbb{P}(\mathbf{e}_q < \tau_0^-)$.

5.2 Resolvent Densities

As an intermediate step to deriving a closed-form expression for the Gerber–Shiu measure, we are interested in computing the so-called *q-resolvent measure* for the Cramér–Lundberg process, X, killed on exiting $[0, \infty)$. Said another way, we are interested in characterising the measure

$$U^{(q)}(a, x, \mathrm{d}y) := \int_0^\infty \mathrm{e}^{-qt} \mathbb{P}_x\big(X_t \in \mathrm{d}y, t < \tau^{[0,a]}\big)\mathrm{d}t, \quad y \in [0, a],$$

where $a > 0$, $q \geq 0$ and

$$\tau^{[0,a]} = \tau_a^+ \wedge \tau_0^-.$$

If, for each $x \in [0, a]$, a density of $U^{(q)}(a, x, \mathrm{d}y)$ exists with respect to Lebesgue measure, then we call it the *resolvent density* and denote it by $u^{(q)}(a, x, y)$. (Note that this density can only be identified Lebesgue almost everywhere.) It turns out that, for each Cramér–Lundberg process, not only does a potential density exist, but we can write it in semi-explicit terms with the help of scale functions. Note, in the statement of the result, it is implicitly understood that $W^{(q)}$ is identically zero on $(0,\infty)$.

Theorem 5.2 *For each $q \geq 0$ and $a > 0$, the measure $U^{(q)}(a,x,\cdot)$ has a density which is equal to*

$$u^{(q)}(a,x,y) = \frac{W^{(q)}(x)W^{(q)}(a-y)}{W^{(q)}(a)} - W^{(q)}(x-y), \quad x,y \in [0,a], \tag{5.2}$$

Lebesgue almost everywhere.

Proof Define, for all $x, y \geq 0$ and $q > 0$,

$$R^{(q)}(x,\mathrm{d}y) = \int_0^\infty \mathrm{e}^{-qt}\mathbb{P}_x\big(X_t \in \mathrm{d}y, t < \tau_0^-\big)\mathrm{d}t.$$

This is the q-resolvent measure for the process X when killed on exiting $[0,\infty)$. Note that, for the same parameter regimes of x, y and q, we can also write

$$R^{(q)}(x,\mathrm{d}y) = \frac{1}{q}\mathbb{P}_x(X_{\mathbf{e}_q} \in \mathrm{d}y, \underline{X}_{\mathbf{e}_q} \geq 0),$$

where, as usual, $\mathbf{e}_q$ is an independent, exponentially distributed random variable with parameter $q > 0$. Appealing to Theorem 5.1, we have that

$$\begin{aligned} R^{(q)}(x,\mathrm{d}y) &= \frac{1}{q}\mathbb{P}\big(x + (X_{\mathbf{e}_q} - \underline{X}_{\mathbf{e}_q}) + \underline{X}_{\mathbf{e}_q} \in \mathrm{d}y, -\underline{X}_{\mathbf{e}_q} \leq x\big) \\ &= \frac{1}{q}\int_{[x-y,x]} \mathbb{P}(-\underline{X}_{\mathbf{e}_q} \in \mathrm{d}z)\mathbb{P}(X_{\mathbf{e}_q} - \underline{X}_{\mathbf{e}_q} \in \mathrm{d}y - x + z). \end{aligned}$$

Note that the delimiter on the integral appears by virtue of the fact that, on the one hand $-\underline{X}_{\mathbf{e}_q} \leq x$ and, on the other hand, if $x + (X_{\mathbf{e}_q} - \underline{X}_{\mathbf{e}_q}) + \underline{X}_{\mathbf{e}_q} \leq y$, then $-\underline{X}_{\mathbf{e}_q} \geq x - y$. By duality, $X_{\mathbf{e}_q} - \underline{X}_{\mathbf{e}_q}$ is equal in distribution to $\overline{X}_{\mathbf{e}_q}$, which itself is exponentially distributed with parameter $\Phi(q)$. In addition, the law of $-\underline{X}_{\mathbf{e}_q}$ has been identified in Corollary 4.2. Using these facts we may write, for $q, x, y \geq 0$,

$$R^{(q)}(x,\mathrm{d}y) = \left\{\int_{[x-y,x]}\left(\frac{1}{\Phi(q)}W^{(q)}(\mathrm{d}z) - W^{(q)}(z)\mathrm{d}z\right)\Phi(q)\mathrm{e}^{-\Phi(q)(y-x+z)}\right\}\mathrm{d}y.$$

In particular, this shows that, for $q, x, y \geq 0$, there exists a density, say $r^{(q)}(x,y)$, for the measure $R^{(q)}(x,\mathrm{d}y)$. Now note that

$$\mathrm{d}\big[\mathrm{e}^{-\Phi(q)z}W^{(q)}(z)\big] = \mathrm{e}^{-\Phi(q)z}\big[W^{(q)}(\mathrm{d}z) - \Phi(q)W^{(q)}(z)\mathrm{d}z\big],$$

and, hence, straightforward integration gives us

$$r^{(q)}(x,y) = \mathrm{e}^{-\Phi(q)y}W^{(q)}(x) - W^{(q)}(x-y), \quad x,y,q \geq 0.$$

Finally, we may use the expression for $r^{(q)}$ to compute the potential density $u^{(q)}$. First, note that, with the help of the strong Markov property,

$$
\begin{aligned}
qU^{(q)}(a,x,\mathrm{d}y) &= \mathbb{P}_x(X_{\mathbf{e}_q} \in \mathrm{d}y, \underline{X}_{\mathbf{e}_q} \geq 0, \overline{X}_{\mathbf{e}_q} \leq a) \\
&= \mathbb{P}_x(X_{\mathbf{e}_q} \in \mathrm{d}y, \underline{X}_{\mathbf{e}_q} \geq 0) \\
&\quad - \mathbb{P}_x(X_{\mathbf{e}_q} \in \mathrm{d}y, \underline{X}_{\mathbf{e}_q} \geq 0, \overline{X}_{\mathbf{e}_q} > a) \\
&= \mathbb{P}_x(X_{\mathbf{e}_q} \in \mathrm{d}y, \underline{X}_{\mathbf{e}_q} \geq 0) \\
&\quad - \mathbb{P}_x\big(X_{\tau^{[0,a]}} = a, \tau^{[0,a]} < \mathbf{e}_q\big)\mathbb{P}_a(X_{\mathbf{e}_q} \in \mathrm{d}y, \underline{X}_{\mathbf{e}_q} \geq 0).
\end{aligned}
$$

The first and third of the three probabilities on the right-hand side above have been computed in the previous paragraph, the second probability is equal to

$$
\mathbb{E}_x\big(\mathrm{e}^{-q\tau_a^+}; \tau_a^+ < \tau_0^-\big) = \frac{W^{(q)}(x)}{W^{(q)}(a)}.
$$

In conclusion, we have that, for $q \geq 0$ and $x \in [0,a]$, $U^{(q)}(a,x,\mathrm{d}y)$ has a density

$$
r^{(q)}(x,y) - \frac{W^{(q)}(x)}{W^{(q)}(a)} r^{(q)}(a,y), \quad y \in [0,a],
$$

which, after a short amount of algebra, can be shown to be equal to the right-hand side of (5.2).

To complete the proof when $q = 0$, one may take limits in (5.2), noting that the measure $U^{(q)}(a,x,\mathrm{d}y)$ is monotone decreasing in q and that the scale function is continuous in q (again, thanks to the Extended Continuity Theorem for Laplace transforms). □

The above proof contains the following corollary for the q-resolvent measure $R^{(q)}(x,\mathrm{d}y)$.

Corollary 5.3 *Fix $q \geq 0$. The q-resolvent measure for X killed on exiting $[0,\infty)$ has a density given by*

$$
r^{(q)}(x,y) = \mathrm{e}^{-\Phi(q)y} W^{(q)}(x) - W^{(q)}(x-y),
$$

for $x, y \geq 0$.

5.3 More on Poisson Integrals

Before putting together the results in the previous section to derive an expression for the Gerber–Shiu measure, we need to briefly return to the issue of Poisson integrals. One can improve upon the result in Theorem 3.4 with a little more work. Recall that

$\{N_t : t \geq 0\}$ is a Poisson process with rate λ and arrival times $\{T_i : i \geq 1\}$, moreover, $\{\xi_i : i \geq 1\}$ are claim sizes.

Theorem 5.4 *Suppose that $f : \mathbb{R} \times [0, \infty) \times (-\infty, 0] \times (0, \infty) \to \mathbb{R}$ is bounded and measurable. Then, for all $t \geq 0$,*

$$\mathbb{E}_x\left(\sum_{i=1}^{N_t} f(T_i, X_{T_i-}, \overline{X}_{T_i-}, \underline{X}_{T_i-}, \xi_i)\right)$$

$$= \lambda \mathbb{E}_x\left(\int_0^t \int_{(0,\infty)} f(s, X_s, \overline{X}_s, \underline{X}_s, u) F(\mathrm{d}u)\mathrm{d}s\right).$$

In particular, by taking limits on both sides of the above equality as $t \to \infty$, thanks to monotonicity, the same result holds when both t and N_t are replaced by ∞.

One can reconstruct the proof of this theorem by following the main steps of Theorem 3.4. In that case, it will be convenient to first show that, in the spirit of Theorem 3.1, the triplet $\{(X_t, \overline{X}_t, \underline{X}_t) : t \geq 0\}$ is a Markov process.

5.4 Gerber–Shiu Measure and Gambler's Ruin

We now have all the tools we need to provide a characterisation of the Gerber–Shiu measure (1.6) in terms of scale functions. In fact, we shall establish an identity for a slightly more general measure. With a slight abuse of notation, for $a > 0$ and $x \in [0, a]$, define

$$K^{(q)}(a, x, \mathrm{d}y, \mathrm{d}z) = \mathbb{E}_x\left[\mathrm{e}^{-q\tau_0^-}; -X_{\tau_0^-} \in \mathrm{d}y, X_{\tau_0^- -} \in \mathrm{d}z, \tau_0^- < \tau_a^+\right],$$

$$z \in [0, a], y \geq 0.$$

Note that the Gerber–Shiu measure, previously called $K^{(q)}(x, \mathrm{d}y, \mathrm{d}z)$, satisfies

$$K^{(q)}(x, \mathrm{d}y, \mathrm{d}z) = K^{(q)}(\infty, x, \mathrm{d}y, \mathrm{d}z),$$

for $y, z \geq 0$. Here is the main result of this chapter.

Theorem 5.5 (Gerber–Shiu measure) *Fix $q \geq 0$ and $a > 0$. Then*

$$K^{(q)}(a, x, \mathrm{d}y, \mathrm{d}z) = \lambda \left\{ \frac{W^{(q)}(x)W^{(q)}(a-z) - W^{(q)}(a)W^{(q)}(x-z)}{W^{(q)}(a)} \right\}$$

$$\times F(z + \mathrm{d}y)\mathrm{d}z,$$

for $x, z \in [0, a]$ and $y \geq 0$. Moreover,

$$K^{(q)}(x, \mathrm{d}y, \mathrm{d}z) = \lambda \left\{ \mathrm{e}^{-\Phi(q)y} W^{(q)}(x) - W^{(q)}(x-y) \right\} F(z + \mathrm{d}y)\mathrm{d}z, \tag{5.3}$$

for $y, z \geq 0$.

Proof Fix $q \geq 0$ and $x \in [0, a]$. For the first identity, it suffices to show that, for all bounded, continuous $f : (0, \infty) \times [0, a] :\to [0, \infty)$,

$$\mathbb{E}_x\left[e^{-q\tau_0^-} f(-X_{\tau_0^-}, X_{\tau_0^- -}); \tau_0^- < \tau_a^+\right]$$
$$= \lambda \int_0^a \int_{(0,\infty)} f(y, z) u^{(q)}(a, x, z) F(z + dy) dz. \tag{5.4}$$

To this end, note that

$$\{\tau_0^- < \tau_a^+\} = \bigcup_{i=1}^{\infty} \{X_{T_i} < 0, \overline{X}_{T_i -} \leq a, \underline{X}_{T_i -} \geq 0\},$$

where the union is taken over disjoint events. It follows with the help of Theorem 5.4 that

$$\mathbb{E}_x\left[e^{-q\tau_0^-} f(-X_{\tau_0^-}, X_{\tau_0^- -}); \tau_0^- < \tau_a^+\right]$$
$$= \mathbb{E}_x\left[\sum_{i=1}^{\infty} \mathbf{1}_{(X_{T_i -} - \xi_i < 0)} \mathbf{1}_{(\overline{X}_{T_i -} \leq a)} \mathbf{1}_{(\underline{X}_{T_i -} \geq 0)} e^{-qT_i} f(-X_{T_i -} + \xi_i, X_{T_i -})\right]$$
$$= \lambda \int_{(0,\infty)} \mathbb{E}_x\left[\int_0^{\infty} \mathbf{1}_{(u > X_{t-})} \mathbf{1}_{(\overline{X}_{t-} \leq a)} \mathbf{1}_{(\underline{X}_{t-} \geq 0)} e^{-qt} f(-X_{t-} + u, X_{t-}) dt\right] F(du)$$
$$= \lambda \int_{(0,\infty)} \mathbb{E}_x\left[\int_0^{\infty} \mathbf{1}_{(u > X_{t-})} \mathbf{1}_{(t < \tau^{[0,a]})} e^{-qt} f(-X_{t-} + u, X_{t-}) dt\right] F(du)$$
$$= \lambda \int_{(0,\infty)} \int_{[0,a]} \int_0^{\infty} \mathbf{1}_{(u > z)} e^{-qt} f(u - z, z) \mathbb{P}_x\left(X_t \in dz, t < \tau^{[0,a]}\right) dt\, F(du). \tag{5.5}$$

Recalling the definition of $U^{(q)}(a, x, dz)$, it follows that

$$\mathbb{E}_x\left[e^{-q\tau_0^-} f(-X_{\tau_0^-}, X_{\tau_0^- -}); \tau_0^- < \tau_a^+\right]$$
$$= \lambda \int_{(0,\infty)} \int_0^a \mathbf{1}_{(u > z)} f(u - z, z) u^{(q)}(a, x, z) dz\, F(du).$$

The right-hand side above is equal to (5.4), after a straightforward application of Fubini's Theorem and a change of variables.

For the second part of the theorem, use monotonicity in a on the left- and right-hand side of (5.5) and take limits as $a \to \infty$. The consequence of this is that we recover the identity

$$\mathbb{E}_x\left[e^{-q\tau_0^-} f(-X_{\tau_0^-}, X_{\tau_0^- -}); \tau_0^- < \infty\right]$$
$$= \lambda \int_0^\infty \int_{(0,\infty)} f(y,z) r^{(q)}(x,z) F(z + dy) dz,$$

for all $x, q \geq 0$, from which (5.3) follows. □

5.5 Comments

In the broader context, Theorem 5.1 is a simple example of one of the several statements that concern the so-called *Wiener–Hopf factorisation* for Lévy processes. See the survey in Bingham (1975), Chapter VI of Bertoin (1996) or Chap. 6 of Kyprianou (2013). The method of analysing the path of the Cramér–Lundberg process (and indeed any random walk) through a sequence of excursions from the minimum was largely popularised by Feller. See for example Chapter XII of Feller (1971). An excursion theoretic treatment of Lévy processes, appealing to an underlying Poissonian structure, can similarly be developed for general Lévy processes. This is a special case of Itô's general theory of excursions for Markov processes, introduced in Itô (1972). Further accounts of general excursion theory can be found in Rogers (1989), Chapter XII of Revuz and Yor (2004) and Chapter IV of Bertoin (1996).

The Gerber–Shiu measure was discussed in the context of the *generalised expected discounted penalty function* in Biffis and Morales (2010). Many of the computations concerning resolvent densities are taken directly from Bertoin (1997), which deals with general spectrally negative Lévy processes. However, older literature dealing with the current setting also exists; see for example Suprun (1976), Korolyuk (1974, 1975a, 1975b), Korolyuk et al. (1976) and Korolyuk and Borovskich (1981).

Chapter 6
Reflection Strategies

Let us now return to the first of the three cases in which we perturb the path of the Cramér–Lundberg process through the payments of dividends. Recall that a reflection (or barrier) strategy consists of paying dividends out of the surplus in such a way that, for a fixed barrier $\mathrm{a} > 0$, any excess of the surplus above this level is instantaneously paid out. The cumulative dividend stream is thus given by $L_t := (\mathrm{a} \vee \overline{X}_t) - \mathrm{a}$, for $t \geq 0$. The resulting trajectory satisfies the dynamics

$$U_t := X_t - L_t = X_t + \mathrm{a} - (\mathrm{a} \vee \overline{X}_t), \quad t \geq 0,$$

with probabilities $\{\mathbb{P}_x : x \in [0, a]\}$. The present value of dividends paid until ruin is thus given by

$$\int_{[0,\varsigma)} \mathrm{e}^{-qt} \mathrm{d}L_t,$$

where $\varsigma = \inf\{t > 0 : U_t < 0\}$.

It is more convenient to start the reflected process from the threshold a. In that case, $\{\mathrm{a} - U_t : t \geq 0\}$, under $\mathbb{P}_{\mathrm{a}}$, is equal in law to $\{Y_t^0 : t \geq 0\}$, under $\mathbb{P}$, where we recall that $Y_t^x := (x \vee \overline{X}_t) - X_t$, $x, t \geq 0$. Moreover, from this point of view, the dividends that are paid out under $\mathbb{P}_{\mathrm{a}}$ are equal in law to the process $\{\overline{X}_t : t \geq 0\}$ under $\mathbb{P}$ and the time of ruin corresponds to

$$\sigma_{\mathrm{a}} = \inf\{t > 0 : Y_t^0 > \mathrm{a}\}.$$

The key object of interest in this chapter, the present value of the dividends paid until ruin under force of interest $q \geq 0$, boils down to the study of the quantity

$$\int_{[0,\sigma_{\mathrm{a}})} \mathrm{e}^{-qt} \mathrm{d}\overline{X}_t \tag{6.1}$$

under $\mathbb{P}$.

A.E. Kyprianou, *Gerber–Shiu Risk Theory*, EAA Series,

DOI 10.1007/978-3-319-02303-8_6,

6.1 Perpetuities

Suppose that $\mathrm{N} = \{\mathrm{N}_t : t \geq 0\}$ is a Poisson process with rate $\alpha > 0$, $\{\zeta_i : i \geq 0\}$ is a sequence of i.i.d. positive random variables with common distribution function G and $b \geq 0$ is a constant. The process

$$\ell_t := bt + \sum_{i=1}^{\mathrm{N}_t} \zeta_i, \quad t \geq 0,$$

is a compound Poisson process with positive jumps and positive drift. Then, in the spirit of (2.3), we can compute its Laplace exponent as follows:

$$\mathbb{E}\big(\mathrm{e}^{-\theta \ell_t}\big) = \mathrm{e}^{-\Lambda(\theta)t}, \quad t, \theta \geq 0,$$

where

$$\Lambda(\theta) = b\theta + \alpha \int_{(0,\infty)} \big(1 - \mathrm{e}^{-\theta x}\big) G(\mathrm{d}x).$$

The key mathematical object that will help us to analyse the present value of dividends paid until ruin is the so-called *perpetuity*[1]

$$\int_0^{\mathbf{e}_p} \mathrm{e}^{-q\ell_t} \mathrm{d}t, \tag{6.2}$$

where $q \geq 0$ and, as usual, $\mathbf{e}_p$ is an independent exponential random variable with rate $p \geq 0$. We have the following main result which characterises its moments.

Theorem 6.1 *For all $n \in \mathbb{N}$,*

$$\mathbb{E}\bigg[\bigg(\int_0^{\mathbf{e}_p} \mathrm{e}^{-q\ell_t} \mathrm{d}t\bigg)^n\bigg] = n! \prod_{k=1}^{n} \frac{1}{p + \Lambda(qk)}.$$

Proof Define, for $t \geq 0$,

$$J_t = \int_t^{\infty} \mathrm{e}^{-q\ell_u} \mathbf{1}_{(u<\mathbf{e}_p)} \mathrm{d}u.$$

Our objective is to compute, for $n \in \mathbb{N}$,

$$\Psi_n := \mathbb{E}\big(J_0^n\big).$$

To this end, note that

$$\frac{\mathrm{d}}{\mathrm{d}t} J_t^n = -n J_t^{n-1} \mathrm{e}^{-q\ell_t} \mathbf{1}_{(t<\mathbf{e}_p)},$$

---

[1] In the definition of a perpetuity, it is more usual to integrate to ∞.

and, hence, we obtain

$$J_0^n - J_t^n = n\int_0^t \mathrm{e}^{-q\ell_u}\mathbf{1}_{(u<\mathbf{e}_p)}J_u^{n-1}\mathrm{d}u. \tag{6.3}$$

Using that $\{\ell_t : t \geq 0\}$ has stationary and independent increments together with the lack of memory property, we have

$$J_t = \mathrm{e}^{-q\ell_t}\mathbf{1}_{(t<\mathbf{e}_p)}J_0^*,$$

where J_0^* is independent of $\{\ell_u : u \leq t\}$ and has the same distribution as J_0. In conclusion, taking expectations in (6.3), we find that

$$\Psi_n\big(1-\mathbb{E}\big(\mathrm{e}^{-nq\ell_t}\mathbf{1}_{(t<\mathbf{e}_p)}\big)\big) = n\Psi_{n-1}\int_0^t \mathbb{E}\big(\mathrm{e}^{-nq\ell_u}\mathbf{1}_{(u<\mathbf{e}_p)}\big)\mathrm{d}u. \tag{6.4}$$

Since

$$\mathbb{E}\big(\mathrm{e}^{-nq\ell_t}\mathbf{1}_{(t<\mathbf{e}_p)}\big) = \exp\big\{-\big(p+\Lambda(nq)\big)t\big\},$$

it follows that

$$\Psi_n = \frac{n}{p+\Lambda(nq)}\Psi_{n-1}.$$

Iterating gives the result. □

Remark 6.2 It is worth noting that Theorem 6.1 and its proof are still valid when we replace the process ℓ by a general subordinator, meaning a non-decreasing Lévy process.

6.2 Decomposing Paths at the Maximum

Let us now look at the reflected process $\{\overline{X}_t - X_t : t \geq 0\}$ and consider how the time it spends in the state zero, as well as the excursions it makes from the state zero, will reveal yet another path decomposition, again incorporating the idea of "Bernoulli trials".

For convenience, write Y_t in place of Y_t^0, $t \geq 0$. Define $S_0 = 0$ and

$$S_0^* = \inf\{t > 0 : Y_t > 0\}.$$

Then continue recursively, so that, for $k \in \mathbb{N}$, on $\{S_{k-1} < \infty\}$,

$$S_k = \inf\big\{t > S_{k-1}^* : Y_t = 0\big\} \quad \text{and} \quad S_k^* = \inf\{t > S_k : Y_t > 0\}$$

and set

$$\chi_k = \overline{X}_{S_{k-1}^*}, \qquad \zeta_k = S_k - S_{k-1}^* \quad \text{and} \quad h_k = \sup_{s\in[S_{k-1}^*, S_k]} Y_s.$$

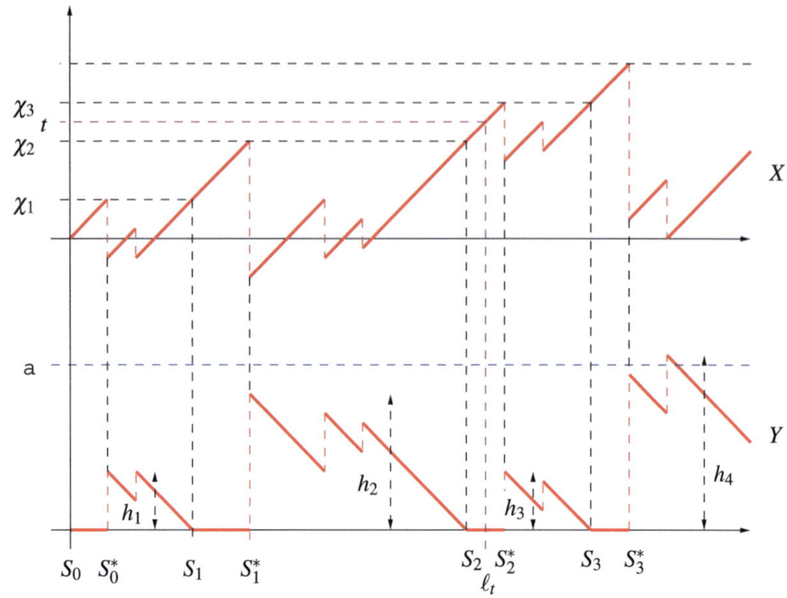

Fig. 6.1 Decomposing the path of X into excursions from its maximum

See Fig. 6.1. Finally, define

$$\upsilon_{\mathtt{a}} = \min\{k \geq 1 : h_k > \mathtt{a}\}.$$

In words, for $k \geq 1$, the intervals $[S_{k-1}, S_{k-1}^*)$ are the successive periods during which the process Y remains at the origin. Equivalently, they are the intervals of time during which the process $\overline{X}$ is increasing. By the lack of memory property, each of these periods, $S_{k-1}^* - S_{k-1}$, is independent and exponentially distributed with the arrival rate λ. The intervals $[S_{k-1}^*, S_k)$ are the successive periods during which the process Y undertakes an excursion from the origin. Equivalently they are the intervals of time during which the process $\overline{X}$ does not increase (and hence X makes an excursion away from $\overline{X}$). The triplets (χ_k, ζ_k, h_k) represent the height of X immediately prior to the k-th excursion, the length of the k-th excursion and the height of k-th excursion, respectively. Moreover, $\upsilon_{\mathtt{a}}$ is the index of the first excursion from the maximum which exceeds height $\mathtt{a}$. Appealing again to the logic behind Bernoulli trials, $\upsilon_{\mathtt{a}}$ is geometrically distributed with parameter

$$r_{\mathtt{a}} = \int_{(0,\infty)} F(\mathrm{d}x)\mathbb{P}_{-x}\left(\tau_{-\mathtt{a}}^- < \tau_0^+\right).$$

Note that, appealing to spatial homogeneity, $r_{\mathtt{a}}$ is the probability that an excursion from the maximum which has initial distance below the previous maximum distributed according to F, reaches the distance $\mathtt{a}$ below the previous maximum before it returns to the level of the previous maximum, thereby completing the excursion. In other words, $r_{\mathtt{a}}$ is the probability that an excursion exceeds $\mathtt{a}$ in height.

Now note that $\overline{X}$ increases if and only if Y is zero. Moreover, when it increases, it does so at a rate $\mathtt{c}$. In other words,

$$\overline{X}_t = \mathtt{c}\int_0^t \mathbf{1}_{(Y_s=0)}\mathrm{d}s. \tag{6.5}$$

With this in mind, we see that the sequence of positive random variables

$$\{\overline{X}_{S_k^*} - \overline{X}_{S_k} : k = 0, \ldots, \upsilon_{\mathtt{a}} - 1\} = \left\{\mathtt{c}\left(S_k^* - S_k\right) : k = 0, \ldots, \upsilon_{\mathtt{a}} - 1\right\}$$

is nothing more than an independent geometric number of independent exponential random variables, each with rate $\lambda/\mathtt{c}$. Therefore, recalling that the sum of an independent geometrically distributed number of i.i.d. exponential random variables is again exponentially distributed, the maximum height reached by X at the moment that it first drops a distance $\mathtt{a}$ below its previous maximum,

$$\overline{X}_{S^*_{\upsilon_{\mathtt{a}}-1}} = \mathtt{c}\sum_{k=0}^{\upsilon_{\mathtt{a}}-1}\left(S_k^* - S_k\right),$$

is exponentially distributed with rate $p_{\mathtt{a}} := \lambda r_{\mathtt{a}}/\mathtt{c}$. In fact, recalling that

$$\chi_k = \overline{X}_{S^*_{k-1}} = \mathtt{c}\sum_{j=0}^{k-1}\left(S_j^* - S_j\right),$$

the triplets

$$\left\{(\chi_k, \zeta_k, h_k) : k = 1, \ldots, \upsilon_{\mathtt{a}}\right\}$$

are the times of arrival, χ_k, and the marks, (ζ_k, h_k), of a marked Poisson process, indexed up to the first of the marks h_k that exceeds $\mathtt{a}$ in value, where the arrival rate is $\lambda/\mathtt{c}$. The marks are distributed according to the joint law

$$H(\mathrm{d}y, \mathrm{d}z) = \int_{(0,\infty)} F(\mathrm{d}x)\mathbb{P}_{-x}\left(\tau_0^+ \in \mathrm{d}y, -\underline{X}_{\tau_0^+} \in \mathrm{d}z\right)\mathbf{1}_{(z\geq x)}, \quad y, z \geq 0.$$

To see why, note, as before, that each excursion begins with a jump below the previous maximum, distributed according to F. The excursion length is therefore the time it takes to reach the previous maximum from this initial position. Moreover, the height of this excursion is also the depth below the previous maximum that X reaches before returning to the level of the previous maximum. Appealing again to spatial homogeneity, the excursion length and height are respectively equivalent to the first passage time τ_0^+ and to the depth below the origin that X reaches before τ_0^+, i.e. $-\underline{X}_{\tau_0^+}$, when X is issued from a position below the origin which is randomised according to F.

Appealing to the Poisson thinning theorem, we can also say that

$$\left\{(\chi_k, \zeta_k, h_k) : k = 1, \ldots, \upsilon_{\mathtt{a}} - 1\right\}$$

is equal in law to the times of arrival and the marks of a Poisson process, say $\mathrm{N}^{\mathrm{a}} = \{\mathrm{N}^{\mathrm{a}}_t : t \geq 0\}$, with arrival rate $\alpha_{\mathrm{a}} = \lambda(1 - r_{\mathrm{a}})/\mathrm{c}$ and mark distribution on $(0, \infty) \times (0, \mathrm{a}]$

$$H_{\mathrm{a}}(\mathrm{d}y, \mathrm{d}z) = \int_{(0,\mathrm{a}]} F(\mathrm{d}x)\mathbb{P}_{-x}\big(\tau_0^+ \in \mathrm{d}y, -\underline{X}_{\tau_0^+} \in \mathrm{d}z | \tau_0^+ < \tau_{-\mathrm{a}}^-\big)\mathbf{1}_{(z \geq x)},$$

when sampled up to an independent and exponentially distributed random time, $\mathbf{e}_{p_{\mathrm{a}}}$. The distribution $H_{\mathrm{a}}(\mathrm{d}y, \mathrm{d}z)$ is the joint law of the excursion length and height conditioned on the excursion height not exceeding a. Indeed, appealing to earlier reasoning, this is equivalent to the law of the pair $(\tau_0^+, -\underline{X}_{\tau_0^+})$ conditioned on the event $\{-\underline{X}_{\tau_0^+} \leq \mathrm{a}\} = \{\tau_0^+ < \tau_{-\mathrm{a}}^-\}$, when X_0 has a random position below the origin, with distance below it distributed according to F. Moreover, recalling that r_{a} is the probability that an excursion has height greater than a, $\lambda(1 - r_{\mathrm{a}})/\mathrm{c}$ is the rate of arrival of excursions with height not exceeding a. Similarly, $\lambda r_{\mathrm{a}}/\mathrm{c}$ is the rate of arrival of excursions with height exceeding a.

Fix $t > 0$ and write

$$\ell_t := \inf\{s > 0 : \overline{X}_s > t\}.$$

In other words, ℓ_t is the amount of time it takes for $\overline{X}$ to climb to the level t. We can split the time horizon $[0, \ell_t]$ into the intervals of time that Y spends at zero (i.e. the time that $\overline{X}$ is climbing) and the intervals of time during which Y undertakes excursions from the origin (i.e. the intervals of time during which $\overline{X}$ is stationary). On the one hand, thanks to the relation (6.5), on the event $\{\ell_t < \infty\}$, the time that Y spends at zero until $\overline{X}$ reaches the level t is given by

$$\int_0^{\ell_t} \mathbf{1}_{(Y_s=0)}\mathrm{d}s = \frac{1}{\mathrm{c}}\overline{X}_{\ell_t} = \frac{t}{\mathrm{c}}.$$

On the other hand, if we further insist that $\{t < \overline{X}_{S^*_{\upsilon_{\mathrm{a}}-1}}\}$ (which necessarily implies the event $\{\ell_t < \infty\}$), then all of the excursions of Y from zero there are less than a in height. Moreover, still on the event $\{t < \overline{X}_{S^*_{\upsilon_{\mathrm{a}}-1}}\}$, their number and lengths are equal in law to $\mathrm{N}^{\mathrm{a}}_t$ and $\{\zeta_i^{\mathrm{a}} : i = 1, \ldots, \mathrm{N}^{\mathrm{a}}_t\}$, respectively, where the ζ_i^{a} are i.i.d. with common distribution given by

$$G_{\mathrm{a}}(\mathrm{d}y) = H_{\mathrm{a}}\big(\mathrm{d}y, [0, \mathrm{a}]\big) = \int_{(0,\mathrm{a}]} F(\mathrm{d}x)\mathbb{P}_{-x}\big(\tau_0^+ \in \mathrm{d}y | \tau_0^+ < \tau_{-\mathrm{a}}^-\big), \quad y \geq 0.$$

In conclusion, we have that, on $\{t < \overline{X}_{S^*_{\upsilon_{\mathrm{a}}-1}}\}$, the time it takes for X to reach height t is equal in law to

$$\ell_t^{\mathrm{a}} := bt + \sum_{i=1}^{\mathrm{N}^{\mathrm{a}}_t} \zeta_i^{\mathrm{a}}, \quad t \geq 0,$$

on the event $\{t < \mathbf{e}_{p_{\mathrm{a}}}\}$, where $b := 1/\mathrm{c}$.

Putting all these pieces together, we develop the expression for the present value of dividends paid until ruin (6.1) for the special case that $x=0$. Indeed, by making a simple change of variables, $t \mapsto \ell_s$, we have

$$\int_{[0,\sigma_{\mathrm{a}})} \mathrm{e}^{-qt}\mathrm{d}\overline{X}_t = \int_0^\infty \mathbf{1}_{(t<S^*_{\upsilon_{\mathrm{a}}-1})}\mathrm{e}^{-qt}\mathrm{d}\overline{X}_t = \int_0^\infty \mathbf{1}_{(u<\overline{X}_{S^*_{\upsilon_{\mathrm{a}}-1}})}\mathrm{e}^{-q\ell_u}\mathrm{d}u$$
$$= \int_0^{\mathbf{e}_{p_{\mathrm{a}}}} \mathrm{e}^{-q\ell^{\mathrm{a}}_u}\mathrm{d}u.$$

We therefore see that, when the surplus process starts at the barrier a, equivalently $x=0$ in (6.1), the present value of dividends until ruin is equal in distribution to the perpetuity (6.2) for appropriate choices of p_{a}, b, α_{a} and G_{a}, as given above.

If we could say a little more about these quantities, then we would be able to develop an expression for the n-th moments of the present value of dividends paid at ruin, at least when the surplus process starts at the barrier, by using Theorem 6.1. That is to say, we would be able to develop further the identity

$$\mathbb{E}\left[\left(\int_{[0,\sigma_{\mathrm{a}})} \mathrm{e}^{-qt}\mathrm{d}\overline{X}_t\right)^n\right] = n!\prod_{k=1}^n \frac{1}{p_{\mathrm{a}}+\Lambda_{\mathrm{a}}(qk)},$$

where $\Lambda_{\mathrm{a}}(\theta) := b\theta + \alpha_{\mathrm{a}}\int_{(0,\mathrm{a}]}(1-\mathrm{e}^{-\theta x})G_{\mathrm{a}}(\mathrm{d}x)$, $\theta \geq 0$. Once again, scale functions come to our assistance.

6.3 Derivative of the Scale Function

Before we can proceed to develop identities using scale functions, we need to say a few words about differentiability, as derivatives of the scale functions will appear in the forthcoming analysis. Recall that, for each $q \geq 0$, the function $W^{(q)}$ is monotone.

Lemma 6.3 *For $x \geq 0$,*

$$W^{(q)}(\mathrm{d}x) = \frac{1}{\mathrm{c}}\delta_0(\mathrm{d}x) + \frac{(\lambda+q)}{\mathrm{c}}W^{(q)}(x)\mathrm{d}x - \left(\frac{\lambda}{\mathrm{c}}\int_{(0,x]} W^{(q)}(x-y)F(\mathrm{d}y)\right)\mathrm{d}x. \tag{6.6}$$

In particular,

$$W^{(q)\prime}_+(x) = \frac{(\lambda+q)}{\mathrm{c}}W^{(q)}(x) - \left(\frac{\lambda}{\mathrm{c}}\int_{(0,x]} W^{(q)}(x-y)F(\mathrm{d}y)\right), \tag{6.7}$$

for $x>0$, where $W^{(q)\prime}_+(x)$ is the right derivative of $W^{(q)}$ at x. Moreover, $W^{(q)}$ is continuously differentiable on $(0,\infty)$ if and only if F has no atoms.

Proof Note that, whereas $W^{(q)}(0-)=0$, we have that, for $q \geq 0$,

$$\begin{aligned} W^{(q)}(0) &= \lim_{\beta\to\infty}\int_0^\infty \beta \mathrm{e}^{-\beta x} W^{(q)}(x)\mathrm{d}x \\ &= \lim_{\beta\to\infty}\frac{\beta}{\psi(\beta)-q} \\ &= \lim_{\beta\to\infty}\frac{1}{\mathrm{c}-\lambda\int_0^\infty \mathrm{e}^{-\beta x}\overline{F}(x)\mathrm{d}x - q/\beta} \\ &= \frac{1}{\mathrm{c}}. \end{aligned}$$

It follows that $W^{(q)}(\mathrm{d}x)$ has an atom at zero of size $1/\mathrm{c}$.

Integrating (4.1) by parts, we have, on the one hand, that

$$\int_{[0,\infty)} \mathrm{e}^{-\beta x} W^{(q)}(\mathrm{d}x) = \int_{(0,\infty)} \mathrm{e}^{-\beta x} W^{(q)}(\mathrm{d}x) + \frac{1}{\mathrm{c}} = \frac{\beta}{\psi(\beta)-q}, \tag{6.8}$$

for $\beta > \Phi(q)$. On the other hand, we also have that

$$\begin{aligned} &\frac{(\lambda+q)}{\mathrm{c}}\int_0^\infty \mathrm{e}^{-\beta x} W^{(q)}(x)\mathrm{d}x - \frac{\lambda}{\mathrm{c}}\int_0^\infty \mathrm{e}^{-\beta x}\int_{(0,x]} W^{(q)}(x-y)F(\mathrm{d}y)\mathrm{d}x \\ &= \frac{1}{\mathrm{c}}\frac{(\lambda+q)}{\psi(\beta)-q} - \frac{1}{\mathrm{c}}\frac{\lambda}{\psi(\beta)-q}\int_{(0,\infty)} \mathrm{e}^{-\beta x}F(\mathrm{d}x) \\ &= \frac{1}{\mathrm{c}}\frac{\lambda\int_{(0,\infty)}(1-\mathrm{e}^{-\beta x})F(\mathrm{d}x)+q}{\psi(\beta)-q} \\ &= \frac{1}{\mathrm{c}}\frac{\mathrm{c}\beta-\psi(\beta)+q}{\psi(\beta)-q} \\ &= \frac{\beta}{\psi(\beta)-q} - \frac{1}{\mathrm{c}}, \end{aligned} \tag{6.9}$$

for $\beta > \Phi(q)$. Comparing the transforms in (6.8) and (6.9), it follows that (6.6) holds. In particular, this implies that $W^{(q)}$ is an absolutely continuous (and hence a continuous) function on $(0,\infty)$.

Thanks to the just-proved fact that $W^{(q)}$ is a continuous function on $(0,\infty)$, inspecting the right-hand side of (6.6), we see that the density of $W^{(q)}$ with respect to Lebesgue measure may be taken as right-continuous. The formula for $W_+^{(q)\prime}(x)$ on $(0,\infty)$ in (6.7) thus follows. Moreover, it is continuous if and only if the convolution is continuous. This is equivalent to requiring that F has no atoms. In that case, on $(0,\infty)$, we have that $W^{(q)}$ is absolutely continuous with a continuous version of its density, thus making it continuously differentiable. □

6.4 Present Value of Dividends Paid Until Ruin

Let us use a completely different technique to examine the first moments of the present value of dividends paid until ruin when the initial value of the surplus is a. This will give us an expression from which we can glean the desired information about the quantities p_{a} and Λ_{a} as per the discussion at the end of Sect. 6.2. Thereafter we can return to the more general problem of the n-th moment of the present value of dividends paid until ruin when the initial value of the surplus is $x \in [0, \mathrm{a}]$.

Lemma 6.4 *For all* $q \geq 0$,

$$\mathbb{E}_{\mathrm{a}}\left[\int_{[0,\varsigma)} \mathrm{e}^{-qt} \mathrm{d}L_t\right] = \mathbb{E}\left[\int_{[0,\sigma_{\mathrm{a}})} \mathrm{e}^{-qt} \mathrm{d}\overline{X}_t\right] = \frac{W^{(q)}(\mathrm{a})}{W_+^{(q)\prime}(\mathrm{a})}. \tag{6.10}$$

Proof The proof works by splitting the integral $\int_0^{\sigma_{\mathrm{a}}} \mathrm{e}^{-qt} \mathrm{d}\overline{X}_t$ at the time of the first jump of X. Note that $\overline{X}$ initially increases at rate c for an independent and exponentially distributed period of time, with rate λ, and then undertakes a jump of size ξ_1 downwards. This jump kick-starts an excursion from the maximum. This excursion will either cause ruin, or its height will remain below the level a and X will return to its previous maximum, during which time no dividends will have been paid and ruin has not occurred. Thereafter, by the strong Markov property, the process we have just described begins again, except that future payments are additionally discounted by the time that has already lapsed. Let

$$V_{\mathrm{a}}^{(1)} = \mathbb{E}\left[\int_{[0,\sigma_{\mathrm{a}})} \mathrm{e}^{-qt} \mathrm{d}\overline{X}_t\right].$$

Following the description given above, we have, with the help of Theorem 4.5,

$$\begin{aligned} V_{\mathrm{a}}^{(1)} &= \mathbb{E}\left[\int_0^{\mathbf{e}_\lambda} \mathrm{e}^{-qt} \mathrm{c} \mathrm{d}u\right] + \mathbb{E}\left[\mathrm{e}^{-q\mathbf{e}_\lambda} \mathbb{E}_{-\xi_1}\left[\mathrm{e}^{-q\tau_0^+} \mathbf{1}_{(\tau_0^+ < \tau_{-\mathrm{a}}^-)}\right] \mathbf{1}_{(\xi_1 \leq \mathrm{a})} V_{\mathrm{a}}^{(1)}\right] \\ &= \frac{\mathrm{c}}{q} \mathbb{E}\left(1 - \mathrm{e}^{-q\mathbf{e}_\lambda}\right) + \mathbb{E}\left(\mathrm{e}^{-q\mathbf{e}_\lambda}\right) \int_{(0,\mathrm{a}]} \frac{W^{(q)}(\mathrm{a}-y)}{W^{(q)}(\mathrm{a})} F(\mathrm{d}y) V_{\mathrm{a}}^{(1)} \\ &= \frac{\mathrm{c}}{\lambda + q} + \frac{\lambda}{\lambda + q} \int_{(0,a]} \frac{W^{(q)}(\mathrm{a}-y)}{W^{(q)}(\mathrm{a})} F(\mathrm{d}y) V_{\mathrm{a}}^{(1)}. \end{aligned} \tag{6.11}$$

From (6.7) we have

$$\frac{\lambda}{\lambda + q} \int_{(0,a]} \frac{W^{(q)}(\mathrm{a}-y)}{W^{(q)}(\mathrm{a})} F(\mathrm{d}y) = 1 - \frac{\mathrm{c}}{\lambda + q} \frac{W_+^{(q)\prime}(\mathrm{a})}{W^{(q)}(\mathrm{a})}.$$

Substituting the above convolution into (6.11) and solving for $V_{\mathrm{a}}^{(1)}$ now gives the statement of the lemma. □

Taking account of the discussion in Sect. 6.2 as well as the conclusion of Theorem 6.1, we see that

$$V_{\mathsf{a}}^{(1)} = \frac{1}{p_{\mathsf{a}} + \Lambda_{\mathsf{a}}(q)} = \frac{W^{(q)}(\mathsf{a})}{W_+^{(q)\prime}(\mathsf{a})}.$$

It follows that, for $n \geq 1$,

$$\begin{aligned} V_{\mathsf{a}}^{(n)} &:= \mathbb{E}_{\mathsf{a}}\left[\left(\int_{[0,\varsigma)} \mathrm{e}^{-qt}\,\mathrm{d}L_t\right)^n\right] \\ &= \mathbb{E}\left[\left(\int_{[0,\sigma_{\mathsf{a}})} \mathrm{e}^{-qt}\,\mathrm{d}\overline{X}_t\right)^n\right] \\ &= n!\prod_{k=1}^{n} \frac{1}{p_{\mathsf{a}} + \Lambda_{\mathsf{a}}(qk)} \\ &= n!\prod_{k=1}^{n} \frac{W^{(qk)}(\mathsf{a})}{W_+^{(qk)\prime}(\mathsf{a})}. \end{aligned}$$

We are almost done. All we need now is to deal with the case that the surplus process starts from any $x \in [0, a]$. To this end, note that, when the Cramér–Lundberg process is issued from $x \in [0, \mathsf{a}]$, dividends are not paid until X reaches the threshold a, providing ruin does not occur beforehand. In that case, future dividend payments are discounted by the time elapsed until reaching the threshold. Hence, by the strong Markov property,

$$\mathbb{E}_x\left[\left(\int_{[0,\varsigma)} \mathrm{e}^{-qt}\,\mathrm{d}L_t\right)^n\right] = \mathbb{E}_x\left[\mathrm{e}^{-qn\tau_{\mathsf{a}}^+}\mathbf{1}_{(\tau_{\mathsf{a}}^+ < \tau_0^-)}\right]V_{\mathsf{a}}^{(n)}.$$

In conclusion, taking account of Theorem 4.5, we have the following main result for this chapter.

Theorem 6.5 *For $q \geq 0$, $n \in \mathbb{N}$ and $x \in [0, \mathsf{a}]$,*

$$\mathbb{E}_x\left[\left(\int_{[0,\varsigma)} \mathrm{e}^{-qt}\,\mathrm{d}L_t\right)^n\right] = n!\frac{W^{(qn)}(x)}{W^{(qn)}(\mathsf{a})}\prod_{k=1}^{n} \frac{W^{(qk)}(\mathsf{a})}{W_+^{(qk)\prime}(\mathsf{a})}.$$

6.5 Comments

As alluded to in the introduction, reflection strategies for Cramér–Lundberg processes emerge naturally from the control problem (1.7). An expression for the expected present value of dividends paid until ruin in terms of (what amounts to) scale functions can be found in the work Gerber (1972). See Eq. (8.9) of that paper. Given that the optimal strategy to (1.7) falls within the class of reflection strategies, one is left with the problem of choosing an optimal value of $\mathtt{a}$ for the threshold. Note from Theorem 6.5 that the expected present value of dividends paid until ruin with reflection at threshold $\mathtt{a}$ is $W^{(q)}(x)/W_+^{(q)\prime}(\mathtt{a})$ when $U_0 = x \leq \mathtt{a}$. Going back to the original work of Gerber (1969, 1972), it is known that $\mathtt{a}$ should be chosen so as to minimise the value of $W_+^{(q)\prime}(\mathtt{a})$. Amongst other literature on this control problem, Loeffen (2008) makes an important step in identifying natural assumptions on the jump distribution F to ensure that $W^{(q)\prime}(\mathtt{a})$ is strictly convex and, hence, can be minimised at a unique value of $\mathtt{a} \geq 0$.

Higher moments of the present value of reflection strategies were first considered in Dickson and Waters (2004) for the case of exponentially distributed jumps. The connection with scale functions was made simultaneously in Renaud and Zhou (2007) and Kyprianou and Palmowski (2007) for the case that X is a general spectrally negative Lévy process. Albrecher and Gerber (2011) extend their result further to the case of stationary upward skip-free Markov processes. The proof we give here is a hybrid version of the proofs found in Dickson and Waters (2004) and Kyprianou and Palmowski (2007). The review article Bertoin and Yor (2005) gives an interesting overview of perpetuities for Lévy processes in general. Decomposing the paths of the Cramér–Lundberg process at the maximum and identifying excursions through a marked Poisson process is an idea that goes back to Greenwood and Pitman (1980) and is based on the earlier-mentioned general theory of excursions for Markov processes due to Itô (1972). Ultimately, however, it is the natural continuous-time analogue of Feller's ideas on the renewal process of maxima (or indeed minima) for random walks. In the context of marked Poisson processes, the thinning theorem is also known as the colouring theorem; see Chap. 5 of Kingman (1993). For further information on the smoothness of scale functions, the reader is referred to Cohen et al. (2013).

Chapter 7
Perturbation-at-Maximum Strategies

In this chapter, we dig a little deeper into the decomposition discussed in Sect. 6.2. In particular, we bring out in more detail the characterisation of the marked Poisson process of excursion heights in terms of scale functions. This will help us to analyse the case where payments are made from the surplus that are proportional to increments of the maximum.

Recall that if $\gamma : [0,\infty) \to [0,\infty)$, then the surplus process, when perturbed at rate γ with respect to its maximum process, yields an aggregate

$$U_t = X_t - \int_{(0,t]} \gamma(\overline{X}_u)\mathrm{d}\overline{X}_u, \quad t \geq 0.$$

We restrict ourselves to the cases mentioned in the introduction: the *heavy-perturbation regime*, for which $\gamma : [0,\infty) \to (1,\infty)$, and the *light-perturbation regime*, for which $\gamma : [0,\infty) \to (0,1)$. The latter may be seen as the result of taxation. The case of a reflection strategy, when $\gamma(x) = \mathbf{1}_{(x \geq \mathrm{a})}$, sits between these two regimes.

7.1 Rehung Excursions

Let us start by noting that, for all $x \geq 0$, under $\mathbb{P}_x$,

$$U_t = A_t - Y_t, \quad t \geq 0, \tag{7.1}$$

where we recall that $Y_t = \overline{X}_t - X_t$ and

$$A_t = \overline{X}_t - \int_{(0,t]} \gamma(\overline{X}_u)\mathrm{d}\overline{X}_u = \bar{\gamma}_x(\overline{X}_t), \quad t \geq 0,$$

with

$$\bar{\gamma}_x(s) := s - \int_x^s \gamma(y)\mathrm{d}y = x + \int_x^s [1-\gamma(y)]\mathrm{d}y, \quad s \geq x.$$

A.E. Kyprianou, *Gerber–Shiu Risk Theory*, EAA Series,
DOI 10.1007/978-3-319-02303-8_7,

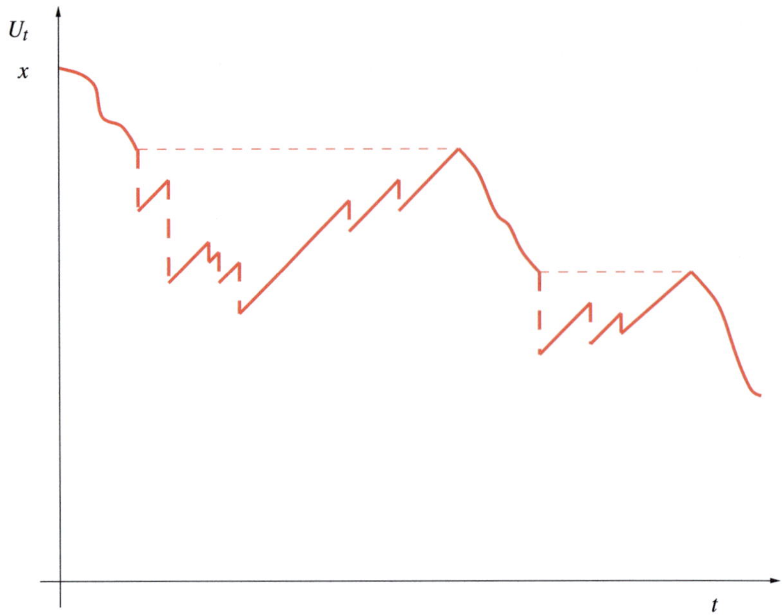

Fig. 7.1 Rehanging excursions from the process A in the heavy-perturbation regime

The last equality shows us that the process $A = \{A_t : t \geq 0\}$ is increasing (resp. decreasing) when γ belongs to the light-perturbation (resp. heavy-perturbation) regime. In essence, monotonicity of A follows from monotonicity of $\overline{X}$ and $\bar{\gamma}_x$. (The latter observation will also be important later as we will use the inverse function $\bar{\gamma}_x^{-1}$, which, accordingly, is well defined in both regimes.) Moreover, A is stationary in value precisely at the times that the process Y is non-zero valued. As a consequence, we may interpret (7.1) as a path decomposition in which excursions of X from its maximum (equivalently excursions of Y away from zero) are "hung" off the trajectory of A during its stationary periods. See Fig. 7.1 for a visual interpretation of this heuristic.

In the light-perturbation regime, since A is an increasing process which is stationary whenever the process Y is non-zero valued, we have that

$$\overleftarrow{U}_t := \sup_{s \leq t} U_s = A_{g_t} = A_t,$$

where $g_t = \sup\{s \leq t : Y_s = 0\}$. Hence, unless it is assumed that

$$\int_x^{\infty} \big(1 - \gamma(s)\big)\mathrm{d}s = \infty, \tag{7.2}$$

in the light-perturbation regime, the perturbed process U will have an almost surely finite global maximum.

In contrast, in the heavy-perturbation regime, when A is decreasing, similar reasoning shows that

$$\overrightarrow{U}_t := \sup_{s \geq t} U_s = A_{d_t} = A_t,$$

where $d_t = \inf\{s > t : Y_s = 0\}$. Hence, the process U is always bounded by its initial value x.

Henceforth, our analysis of the process U will centre around further analysis of the excursions of X from its maximum $\overline{X}$. In particular, their representation through a marked Poisson process, such as we saw in Sect. 6.2, will prove to be extremely useful. In the next section, we shall revisit this marked Poisson process and look at a more detailed characterisation of its parameters in terms of scale functions.

7.2 Marked Poisson Process Revisited

Recall from the previous chapter that we write $Y_t = \overline{X}_t - X_t$, for $t \geq 0$. Moreover, we recursively defined $S_0 = 0$,

$$S_0^* = \inf\{t > 0 : Y_t > 0\}$$

and, for $k \in \mathbb{N}$, on $\{S_{k-1} < \infty\}$,

$$S_k = \inf\left\{t > S_{k-1}^* : Y_t = 0\right\} \quad \text{and} \quad S_k^* = \inf\{t > S_k : Y_t > 0\}.$$

The k-th excursion from the maximum occurs over the time intervals $[S_{k-1}^*, S_k]$, and the height of the excursion was denoted by

$$h_k = \sup_{s \in [S_{k-1}^*, S_k]} Y_s.$$

Now let

$$\upsilon_\infty = \min\{k \geq 1 : S_k = \infty\},$$

the index of the first excursion which is infinite in length. Note, by Lemma 3.10, that if $\psi'(0+) \geq 0$, then $\upsilon_\infty = \infty$ almost surely. Moreover, if $\psi'(0+) < \infty$, then υ_∞ is geometrically distributed with parameter

$$r_\infty = \int_{(0,\infty)} F(\mathrm{d}x)\mathbb{P}_{-x}\left(\tau_0^+ = \infty\right).$$

In both cases, it is clear that we may equivalently define

$$\upsilon_\infty = \min\{k \geq 1 : h_k = \infty\}.$$

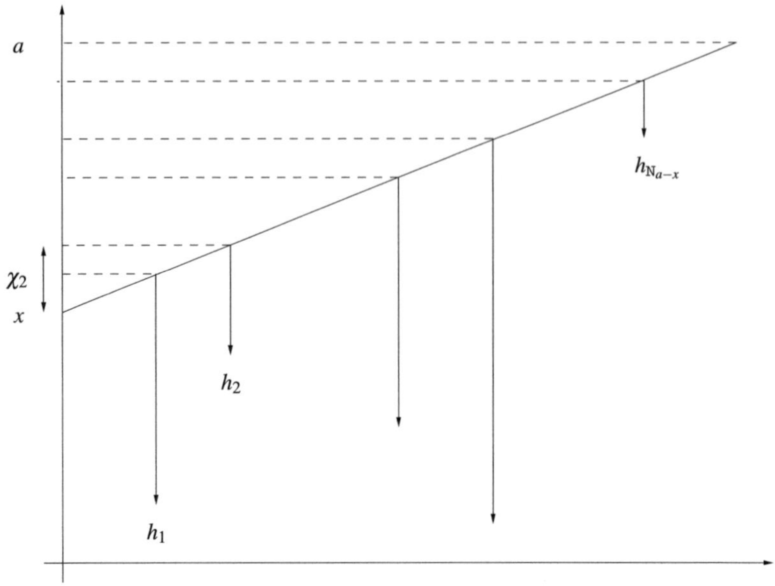

Fig. 7.2 Excursion heights as the process X climbs from x to a

Under $\mathbb{P}_x$, define in a similar manner to the previous chapter $\chi_k = \overline{X}_{S^*_{k-1}} - x$, where we now allow the index k to run from 1 to υ_∞. In the spirit of the reasoning in Sect. 6.2, we also note that

$$\left\{(\chi_k, h_k) : k = 1, \ldots, \upsilon_\infty\right\}$$

is equal to times of arrival and the marks of a marked Poisson process, say $\mathrm{N} = \{\mathrm{N}_t : t \geq 0\}$, with rate λ/c up until the first mark which is infinite in value. Moreover, marks are i.i.d. (and independent of N) with common distribution

$$Q(\mathrm{d}y) := \int_{(0,\infty)} F(\mathrm{d}x)\mathbb{P}_{-x}(-\underline{X}_{\tau_0^+} \in \mathrm{d}y), \quad y \in [0, \infty].$$

Let us return to the event $\{\tau_a^+ < \tau_0^-\}$, for $a > 0$. Note that, for $0 \leq x \leq a$,

$$\mathbb{P}_x\left(\tau_a^+ < \tau_0^-\right) = \mathbb{P}(h_k \leq x + \chi_k \text{ for } k = 1, \ldots, \mathrm{N}_{a-x}).$$

Figure 7.2 gives a visual explanation of this equality.

Classical theory for Poisson processes tells us that, conditional on $\{\mathrm{N}_{a-x} = n\}$, the arrival times $\{\chi_1, \ldots, \chi_n\}$ are equal in law to an ordered i.i.d. sample of uniformly distributed points on $[0, a-x]$. Then,

$$\begin{aligned}
&\mathbb{P}(h_k \leq x + \chi_k \text{ for } k = 1, \ldots, \mathrm{N}_{a-x}) \\
&= \sum_{n=0}^{\infty} \mathrm{e}^{-\frac{\lambda}{\mathrm{c}}(a-x)} \frac{1}{n!} \left(\frac{\lambda(a-x)}{\mathrm{c}}\right)^n \left(\int_0^{a-x} Q(x+t)\frac{1}{(a-x)}\mathrm{d}t\right)^n
\end{aligned}$$

$$= \exp\left\{\frac{\lambda}{c}\int_0^{a-x}(Q(x+t)-1)\mathrm{d}t\right\}$$

$$= \exp\left\{-\frac{\lambda}{c}\int_x^a \overline{Q}(y)\mathrm{d}y\right\},$$

where $\overline{Q}(y) = 1 - Q(y)$, for $y \geq 0$.

It was proved earlier that $\mathbb{P}_x(\tau_a^+ < \tau_0^-) = W(x)/W(a)$, where W is the scale function associated to X. This leads us to the identity

$$\frac{W(x)}{W(a)} = \exp\left\{-\frac{\lambda}{c}\int_x^a \overline{Q}(y)\mathrm{d}y\right\}. \tag{7.3}$$

Note that this confirms the conclusion of Lemma 6.3 that W is almost everywhere differentiable. Taking account of the above discussion and recalling that W'_+ is the right derivative of W on $(0, \infty)$, we arrive at the following important result.

Theorem 7.1 *For all $x > 0$,*

$$\frac{\lambda}{c}\overline{Q}(x) = \frac{W'_+(x)}{W(x)}. \tag{7.4}$$

Later on in this chapter, when using this result, the quantity $\overline{Q}$ will always appear in the context of a Lebesgue integral. In that case, it suffices to write W'/W on the right-hand side of (7.4), without needing to refer to the right derivative of W.

7.3 Gambler's Ruin for the Perturbed Process

Let

$$T_0^- := \inf\{t > 0 : U_t < 0\}.$$

In the light-perturbation regime, we may write for all values b in the range of $\bar{\gamma}_x$,

$$\tau^+_{\bar{\gamma}_x^{-1}(b)} = T_b^+, \tag{7.5}$$

where

$$T_b^+ = \inf\{t > 0 : U_t > b\}.$$

Theorem 7.2 *Fix $x > 0$ and assume (7.2) in the case of the light-perturbation regime. In the case of the heavy-perturbation regime, define*

$$s^*(x) = \inf\{s \geq x : \bar{\gamma}_x(s) < 0\}.$$

Then, for any $q \geq 0$, *and* $0 \leq x \leq a$ *in the case of light perturbation, resp.* $0 \leq x \leq a < s^*(x)$ *in the case of heavy perturbation, we have*

$$\mathbb{E}_x\left[\mathrm{e}^{-q\tau_a^+}\mathbf{1}_{(\tau_a^+ < T_0^-)}\right] = \exp\left(-\int_x^a \frac{W^{(q)\prime}(\bar{\gamma}_x(s))}{W^{(q)}(\bar{\gamma}_x(s))}\mathrm{d}s\right). \tag{7.6}$$

Before moving to the proof of this result, let us remark that, taking account of the equivalence (7.5) in the light-perturbation regime, (7.6) can also be written as

$$\mathbb{E}_x\left[\mathrm{e}^{-qT_a^+}\mathbf{1}_{(T_a^+ < T_0^-)}\right] = \exp\left(-\int_x^{\bar{\gamma}_x^{-1}(a)} \frac{W^{(q)\prime}(\bar{\gamma}_x(s))}{W^{(q)}(\bar{\gamma}_x(s))}\mathrm{d}s\right).$$

Proof of Theorem 7.2 The proof does not distinguish between the two different regimes of light and heavy perturbation. All that is required in what follows is that $\bar{\gamma}_x^{-1}(a) < \infty$. Thereafter, the proof needs little more than to recycle a number of existing computations we have already seen.

Using the Esscher transform and appealing to similar reasoning as in the proof of Theorem 7.1, we have

$$\begin{aligned}
&\mathbb{E}_x\left[\mathrm{e}^{-q\tau_a^+}\mathbf{1}_{(\tau_a^+ < T_0^-)}\right] \\
&= \mathrm{e}^{-(a-x)\Phi(q)}\mathbb{P}_x^{\Phi(q)}\left(\tau_a^+ < T_0^-\right) \\
&= \mathrm{e}^{-(a-x)\Phi(q)}\mathbb{P}^{\Phi(q)}\left(h_k \leq \bar{\gamma}_x(x+\chi_k) \text{ for } k = 1, \ldots, \mathrm{N}_{a-x}\right) \\
&= \mathrm{e}^{-(a-x)\Phi(q)}\sum_{n=0}^{\infty}\mathrm{e}^{-\frac{\lambda}{\mathrm{c}}(a-x)}\frac{1}{n!}\left(\frac{\lambda(a-x)}{\mathrm{c}}\right)^n \\
&\quad\times\left(\int_0^{a-x} Q_{\Phi(q)}\left(\bar{\gamma}_x(x+t)\right)\frac{1}{(a-x)}\mathrm{d}t\right)^n \\
&= \exp\left(-\int_0^{a-x} \Phi(q) + \frac{\lambda}{\mathrm{c}}\overline{Q}_{\Phi(q)}\left(\bar{\gamma}_x(x+t)\right)\mathrm{d}t\right),
\end{aligned} \tag{7.7}$$

where $Q_{\Phi(q)}$ plays the role of the quantity Q under the measure $\mathbb{P}^{\Phi(q)}$ and $\overline{Q}_{\Phi(q)} = 1 - Q_{\Phi(q)}$. In particular, recalling the conclusion of Theorems 2.3 and 7.1, we have, for Lebesgue almost every $x \geq 0$,

$$\frac{\lambda}{\mathrm{c}}\overline{Q}_{\Phi(q)}(x) = \frac{W_{\Phi(q)}'(x)}{W_{\Phi(q)}(x)}.$$

Moreover, recalling from the definition (4.4) that $W^{(q)}(x) = \mathrm{e}^{\Phi(q)x}W_{\Phi(q)}(x)$, $x \geq 0$, we have that, for Lebesgue almost every $x \geq 0$,

$$\frac{W^{(q)\prime}(x)}{W^{(q)}(x)} = \Phi(q) + \frac{W_{\Phi(q)}'(x)}{W_{\Phi(q)}(x)}.$$

Putting the pieces together in (7.7) produces the required identity. □

Theorem 7.2 motivates some interesting observations concerning the event of ruin, $\{T_0^- < \infty\}$. First, suppose that we are in the heavy-perturbation regime and $s^*(x) < \infty$. In that case

$$\mathbb{P}_x\big(T_0^- < \infty\big) \geq \mathbb{P}_x\big(\tau_{s^*(x)}^+ < \infty\big) \vee \mathbb{P}_x\big(\tau_0^- < \infty\big).$$

Indeed, on the event $\{\tau_{s^*(x)}^+ < \infty\}$, we have $\overline{X}_{\tau_{s^*(x)}^+} - X_{\tau_{s^*(x)}^+} = 0$ and hence $U_{\tau_{s^*(x)}^+} = A_{\tau_{s^*(x)}^+} = \bar{\gamma}_x(s^*(x)) = 0$. Moreover, since $U_t \leq X_t$ for all $t \geq 0$, it follows that $\{\tau_0^- < \infty\} \subseteq \{T_0^- < \infty\}$. In the event that $\limsup_{t\to\infty} X_t = \infty$ almost surely, we have $\mathbb{P}_x(\tau_{s^*(x)}^+ < \infty) = 1$. Otherwise, it follows that $\mathbb{P}_x(\tau_0^- < \infty) = 1$. Either way, $\mathbb{P}_x(T_0^- < \infty) = 1$.

Remaining in the heavy-perturbation regime, suppose that $s^*(x) = \infty$. Then from (7.6), by taking limits as $a \to \infty$, we get an expression for the ruin probability,

$$\mathbb{P}_x\big(T_0^- < \infty\big) = 1 - \exp\bigg(-\int_x^\infty \frac{W'(\bar{\gamma}_x(s))}{W(\bar{\gamma}_x(s))}\mathrm{d}s\bigg). \tag{7.8}$$

However, the right-hand side above turns out to be equal to 1. Recalling that $W'(x)/W(x) = \lambda\overline{Q}(x)/c$ for Lebesgue almost every $x > 0$, since $\overline{Q}(x)$ is non-increasing on $(0,\infty)$ and $\bar{\gamma}_x(s) \leq x$ for all $s \geq 0$, the claim follows.

Finally, in the light-perturbation regime, where necessarily $s^*(x) = \infty$, the reasoning that leads to (7.8) still applies, from which one may deduce that this probability need not be unity. Indeed, suppose that we take the function γ to be simply constant in value, also denoted by $\gamma \in (0,1)$, and $\psi'(0+) > 0$. In that case, $\bar{\gamma}_x(s) = (s-x)(1-\gamma)+x$, and hence, noting that $\mathrm{d}[\log W(x)]/\mathrm{d}x = W'(x)/W(x)$ Lebesgue almost everywhere on $(0,\infty)$, after a change of variables, we have

$$\mathbb{P}_x\big(T_0^- < \infty\big) = 1 - \exp\bigg(-\frac{1}{(1-\gamma)}\int_x^\infty \frac{W'(u)}{W(u)}\mathrm{d}u\bigg) \tag{7.9}$$

$$= 1 - \big(\psi'(0+)W(x)\big)^{1/(1-\gamma)}, \tag{7.10}$$

for $x \geq 0$, where we have used, from (4.2), that $W(\infty) = 1/\psi'(0+)$.

7.4 Continuous Ruin with Heavy Perturbation

Although the perturbed process is almost surely ruined in the heavy-perturbation regime, it is interesting to note that, unlike a Cramér–Lundberg process, there are two different ways to become ruined. The first, i.e. by a jump downwards, is a property inherited from the underlying process, X. The other way of becoming ruined, which we refer to as *continuous ruin*, is the result of continuously passing the origin at the moment in time that an increment in $\overline{X}$ brings U along the curve $\bar{\gamma}_x$ just as it intersects the origin. Said another way, continuous ruin corresponds to the event that $\{\tau^+_{s^*(x)} = T_0^-\}$, in which case, as remarked upon above, $U_{\tau^+_{s^*(x)}} = 0$. This can only happen with positive probability if $s^*(x) < \infty$.

The following result is a corollary to Theorem 7.2 on account of the fact that its proof is identical, albeit that one replaces a by $s^*(x)$.

Corollary 7.3 *Fix $x > 0$, and suppose that $\gamma : [0,\infty) \to (1,\infty)$ such that $s^*(x) < \infty$. Then, for all $q \geq 0$,*

$$\mathbb{E}_x\left[\mathrm{e}^{-qT_0^-}1_{\{T_0^-=\tau^+_{s^*(x)}\}}\right] = \exp\left(-\int_x^{s^*(x)} \frac{W^{(q)\prime}(\bar{\gamma}_x(s))}{W^{(q)}(\bar{\gamma}_x(s))}\mathrm{d}s\right).$$

If we take the case that $\gamma(s)$ is a constant valued in $(1,\infty)$, again denoted by γ, then we may simplify the formula in the above corollary. Indeed, we have $\bar{\gamma}_x(s) = x - (s-x)(\gamma-1)$ and hence $s^*(x) = \gamma x/(\gamma-1)$. Moreover, a straightforward computation, in a similar vein to the computation in (7.10), gives us

$$\mathbb{P}_x(\text{continuous ruin}) = \exp\left(-\frac{1}{(\gamma-1)}\int_0^x \frac{W'(u)}{W(u)}\mathrm{d}u\right) = \left(\frac{1}{\mathrm{c}W(x)}\right)^{1/(\gamma-1)},$$

for $x \geq 0$, where we have used, from Lemma 6.3, that $W(0) = 1/\mathrm{c}$.

7.5 Expected Present Value of Tax at Ruin

In the spirit of the Gerber–Shiu-type results presented in the previous sections, our final theorem for perturbed processes (with either light or heavy perturbation) considers the expected present value of the payout until ruin. In the case of light perturbation, this corresponds to the expected present value of tax paid until ruin.

Theorem 7.4 *Fix $x \geq 0$ and assume (7.2) in the case of the light-perturbation regime. Then, for $q \geq 0$,*

$$\mathbb{E}_x\left[\int_0^{T_0^-} \mathrm{e}^{-qu}\gamma(\overline{X}_u)\mathrm{d}\overline{X}_u\right] = \int_x^{s^*(x)} \exp\left(-\int_x^t \frac{W^{(q)\prime}(\bar{\gamma}_x(s))}{W^{(q)}(\bar{\gamma}_x(s))}\mathrm{d}s\right)\gamma(t)\mathrm{d}t.$$

Proof Appealing to a straightforward change of variables and Fubini's Theorem, noting in particular that $\tau_t^+ = \inf\{s > 0 : \overline{X}_s > t\}$, we have

$$\begin{aligned}\mathbb{E}_x\left[\int_0^{T_0^-} \mathrm{e}^{-qu}\gamma(\overline{X}_u)\mathrm{d}\overline{X}_u\right] &= \mathbb{E}_x\left[\int_0^{s^*(x)} \mathbf{1}_{(u<T_0^-)}\mathrm{e}^{-qu}\gamma(\overline{X}_u)\mathrm{d}\overline{X}_u\right] \\ &= \mathbb{E}_x\left[\int_x^{s^*(x)} \mathbf{1}_{(\tau_t^+<T_0^-)}\mathrm{e}^{-q\tau_t^+}\gamma(t)\mathrm{d}t\right] \\ &= \int_x^{s^*(x)} \mathbb{E}_x\left[\mathrm{e}^{-q\tau_t^+}\mathbf{1}_{(\tau_t^+<T_0^-)}\right]\gamma(t)\mathrm{d}t.\end{aligned}$$

The proof is completed by taking advantage of the identity in (7.6). □

We again get a simplification of this formula in the case that we take the function $\gamma(s)$ to be a constant, for either the light-perturbation or heavy-perturbation regime. For example when $\gamma(s)$ is equal to the constant $\gamma \in (0,1)$, one gets, for $x, q \geq 0$,

$$\mathbb{E}_x\left[\int_0^{T_0^-} \mathrm{e}^{-qu}\gamma(\overline{X}_u)\mathrm{d}\overline{X}_u\right] = \frac{\gamma}{1-\gamma}\int_x^\infty \left(\frac{W^{(q)}(x)}{W^{(q)}(z)}\right)^{1/(1-\gamma)} \mathrm{d}z.$$

7.6 Comments

The well-known joint law of the first $n \geq 0$ arrival times of a Poisson process conditioned to have n arrivals up to a fixed time can be found, for example, in Sect. 2.4 of Kingman (1993). The representation of the scale function in the form (7.3) is lifted from Theorem 8 of Chapter VII in Bertoin (1996). This representation and the observation that excursions are rehung from the process A form a key part of the analysis in Albrecher et al. (2008), Kyprianou and Zhou (2009) and Kyprianou and Ott (2012), in increasing degrees of generality.

Chapter 8
Refraction Strategies

Let us return to the case of the refracted Cramér–Lundberg process, that is, the solution to the stochastic differential equation (SDE)

$$\mathrm{d}Z_t = \mathrm{d}X_t - \delta \mathbf{1}_{(Z_t > \mathrm{b})}\mathrm{d}t, \quad t \geq 0,$$

also written as

$$Z_t = X_t - \delta \int_0^t \mathbf{1}_{(Z_s > \mathrm{b})}\mathrm{d}s, \quad t \geq 0, \tag{8.1}$$

for some threshold $\mathrm{b} \geq 0$. We shall charge ourselves with the task of providing identities for the probability of ruin as well as the expected present value of dividends paid until ruin. As we have seen earlier, for the case of a Cramér–Lundberg process, it turns out to be convenient to first derive an identity for the resolvent measure of the refracted process until first exiting a finite interval. It turns out that all identities can be written in terms of two scale functions for two different Cramér–Lundberg processes. As one might expect these identities are somewhat more complicated.

8.1 Pathwise Existence and Uniqueness

Before we can look at functionals which pertain to Gerber–Shiu theory, we are confronted with the more pressing issue of whether a solution to this SDE (8.1) exists. As the reader might already suspect, problems may occur when $\delta \geq \mathrm{c}$, as, in that case, when dividends are paid, it is at a higher rate than the premiums collected. We therefore assume throughout that

$$0 < \delta < \mathrm{c}.$$

Theorem 8.1 *The SDE* (8.1) *has a unique* pathwise[1] *solution.*

[1] We can understand the words "pathwise solution" here to mean a solution which is constructed directly from the path of the driving process X. More commonly, such solutions are called *strong solutions*.

A.E. Kyprianou, *Gerber–Shiu Risk Theory*, EAA Series,
DOI 10.1007/978-3-319-02303-8_8,

Proof Define the times $T_n^{\uparrow}$ and $T_n^{\downarrow}$ recursively as follows. We set $T_0^{\downarrow}=0$ and, for $n=1,2,\ldots$, on the events $\{T_{n-1}^{\downarrow}<\infty\}$ and $\{T_n^{\uparrow}<\infty\}$ respectively, set

$$T_n^{\uparrow}=\inf\left\{t>T_{n-1}^{\downarrow}:X_t-\delta\sum_{i=1}^{n-1}(T_i^{\downarrow}-T_i^{\uparrow})>\mathrm{b}\right\},$$

$$T_n^{\downarrow}=\inf\left\{t>T_n^{\uparrow}:X_t-\delta\sum_{i=1}^{n-1}(T_i^{\downarrow}-T_i^{\uparrow})-\delta\big(t-T_n^{\uparrow}\big)<\mathrm{b}\right\}.$$

The difference between the two consecutive times $T_n^{\uparrow}$ and $T_n^{\downarrow}$ is strictly positive. Moreover, $\lim_{n\uparrow\infty}T_n^{\uparrow}=\lim_{n\uparrow\infty}T_n^{\downarrow}=\infty$, almost surely. Now we construct a solution to (8.1), $Z=\{Z_t:t\geq 0\}$, as follows. The process is issued from $X_0=x$ and

$$Z_t=\begin{cases}X_t-\delta\sum_{i=1}^{n}(T_i^{\downarrow}-T_i^{\uparrow}), & \text{for } t\in[T_n^{\downarrow},T_{n+1}^{\uparrow}) \text{ and } n\geq 0,\\ X_t-\delta\sum_{i=1}^{n-1}(T_i^{\downarrow}-T_i^{\uparrow})-\delta(t-T_n^{\uparrow}), & \text{for } t\in[T_n^{\uparrow},T_n^{\downarrow}) \text{ and } n\geq 1.\end{cases}$$

Note that for $n=1,2,\ldots$, on the events $\{T_{n-1}^{\downarrow}<\infty\}$ and $\{T_n^{\uparrow}<\infty\}$ respectively, the times $T_n^{\uparrow}$ and $T_n^{\downarrow}$, can then be identified as

$$T_n^{\uparrow}=\inf\big\{t>T_{n-1}^{\downarrow}:Z_t>\mathrm{b}\big\},\qquad T_n^{\downarrow}=\inf\big\{t>T_n^{\uparrow}:Z_t<\mathrm{b}\big\}.$$

Hence

$$Z_t=X_t-\delta\int_0^t\mathbf{1}_{(Z_s>\mathrm{b})}\mathrm{d}s,\quad t\geq 0,$$

thereby proving the existence of a pathwise solution to (8.1).

For uniqueness of this solution, suppose that $\{Z_t^{(1)}:t\geq 0\}$ and $\{Z_t^{(2)}:t\geq 0\}$ are two pathwise solutions to (8.1). Then, writing

$$\Delta_t=Z_t^{(1)}-Z_t^{(2)}=-\delta\int_0^t(\mathbf{1}_{(Z_s^{(1)}>\mathrm{b})}-\mathbf{1}_{(Z_s^{(2)}>\mathrm{b})})\mathrm{d}s,$$

it follows from integration by parts that

$$\Delta_t^2=-2\delta\int_0^t\Delta_s(\mathbf{1}_{(Z_s^{(1)}>\mathrm{b})}-\mathbf{1}_{(Z_s^{(2)}>\mathrm{b})})\mathrm{d}s.$$

Thanks to the fact that $\mathbf{1}_{(x>\mathrm{b})}$ is an increasing function, it follows from the above representation, that, for all $t\geq 0$, $\Delta_t^2\leq 0$ and hence $\Delta_t=0$ almost surely, thereby proving uniqueness of pathwise solutions to (8.1). □

Let us momentarily return to the reason why Z is referred to as a refracted Lévy process. A simple sketch of a realisation of the path of Z (see for example Fig. 8.1) gives the impression that the trajectory of Z "refracts" each time it passes continu-

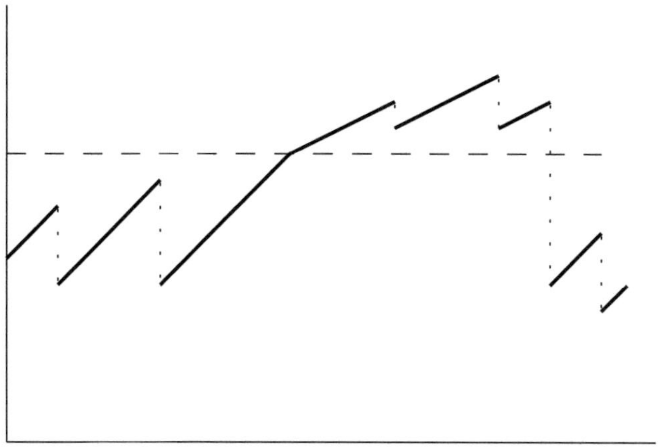

Fig. 8.1 A sample path of Z when the driving Lévy process is a Cramér–Lundberg process. Its trajectory "refracts" as it passes continuously above the *horizontal dashed line* at level b

ously from $(-\infty, \mathrm{b}]$ into (b, ∞), much as a beam of light does when passing from one medium to another.

The construction of the unique pathwise solution described above clearly shows that Z is adapted to the natural filtration $\mathbb{F} = \{\mathcal{F}_t : t \geq 0\}$ of X. Conversely, since, for all $t \geq 0$, $X_t = Z_t + \delta \int_0^t \mathbf{1}_{(Z_s > \mathrm{b})} \mathrm{d}s$, it is also clear that X is adapted to the natural filtration of Z. We can use this observation to reason that Z is a strong Markov process.

To this end, suppose that T is a stopping time with respect to $\mathbb{F}$. Then define a process $\widehat{Z}$ whose dynamics are those of $\{Z_t : t \leq T\}$ issued from $x \in \mathbb{R}$ and, given $\mathcal{F}_T$, on the event that $\{T < \infty\}$, it continues to evolve on the time interval $[T, \infty)$ as the unique solution, say $\widetilde{Z}$, to (8.1) but driven by the Lévy process $\widetilde{X} = \{X_{T+s} - X_T : s \geq 0\}$ and issued from Z_T. Note that, by construction, on $\{T < \infty\}$, the dependence of $\{\widehat{Z}_t : t \geq T\}$ on $\{\widehat{Z}_t : t \leq T\}$ occurs only through the value $\widehat{Z}_T = Z_T$. Note also that for $t > 0$,

$$
\begin{aligned}
\widehat{Z}_{T+t} &= \widetilde{Z}_t \\
&= \widehat{Z}_T + \widetilde{X}_t - \delta \int_0^t \mathbf{1}_{(\widetilde{Z}_s > \mathrm{b})} \mathrm{d}s \\
&= x + X_T - \delta \int_0^T \mathbf{1}_{(Z_s > \mathrm{b})} \mathrm{d}s + (X_{T+t} - X_T) - \delta \int_0^t \mathbf{1}_{(\widehat{Z}_{T+s} > \mathrm{b})} \mathrm{d}s \\
&= x + X_{T+t} - \delta \int_0^{T+t} \mathbf{1}_{(\widehat{Z}_s > \mathrm{b})} \mathrm{d}s,
\end{aligned}
$$

thereby showing that $\widehat{Z}$ solves (8.1) with $\widehat{Z}_0 = x$. Since (8.1) has a unique pathwise solution, this solution must be $\widehat{Z}$ and therefore possesses the strong Markov property.

8.2 Gambler's Ruin and Resolvent Density

Let us now introduce the stopping times for Z

$$\kappa_a^+ := \inf\{t > 0 : Z_t > a\} \quad \text{and} \quad \kappa_0^- := \inf\{t > 0 : Z_t < 0\},$$

where $a > 0$. We are interested in studying the ruin probability

$$\mathbb{P}_x\big(\kappa_0^- < \infty\big), \tag{8.2}$$

as well as the expected present value of dividends paid until ruin,

$$\delta \mathbb{E}_x\left(\int_0^{\kappa_0^-} \mathrm{e}^{-qt} \mathbf{1}_{(Z_t > \mathrm{b})} \mathrm{d}t\right). \tag{8.3}$$

Not unlike our treatment of the analogous object for the Cramér–Lundberg process, it turns out to be more convenient to first study the seemingly more complex two-sided exit problem. To this end, let $X^\delta = \{X_t^\delta : t \geq 0\}$, where $X_t^\delta = X_t - \delta t$ and denote by P_x the law of the process X^δ when issued from x (with E_x as the associated expectation operator). For each $q \geq 0$, $W^{(q)}$ and $Z^{(q)}$ denote, as usual, the q-scale functions associated with X. We shall write $\mathbb{W}^{(q)}$ for the q-scale function associated with X^δ. For convenience, we will write

$$w^{(q)}(x; y) = W^{(q)}(x - y) + \delta \mathbf{1}_{(x \geq b)} \int_{\mathrm{b}}^x \mathbb{W}^{(q)}(x - z) W^{(q)\prime}(z - y) \mathrm{d}z,$$

for $x, y \in \mathbb{R}$ and $q \geq 0$. We have two main results concerning the gambler's ruin problem, from which more can be said about the quantities (8.2) and (8.3).

Theorem 8.2 *For $q \geq 0$ and $0 \leq x, \mathrm{b} \leq a$, we have*

$$\mathbb{E}_x\big(\mathrm{e}^{-q\kappa_a^+} \mathbf{1}_{(\kappa_a^+ < \kappa_0^-)}\big) = \frac{w^{(q)}(x; 0)}{w^{(q)}(a; 0)}. \tag{8.4}$$

Theorem 8.3 *For $q \geq 0$ and $0 \leq x, y, \mathrm{b} \leq a$,*

$$\begin{aligned} &\int_0^\infty \mathrm{e}^{-qt} \mathbb{P}_x\big(Z_t \in \mathrm{d}y, t < \kappa_0^- \wedge \kappa_a^+\big) \mathrm{d}t \\ &= \mathbf{1}_{(y \in [\mathrm{b}, a])} \left\{ \frac{w^{(q)}(x; 0)}{w^{(q)}(a; 0)} \mathbb{W}^{(q)}(a - y) - \mathbb{W}^{(q)}(x - y) \right\} \mathrm{d}y \\ &\quad + \mathbf{1}_{(y \in [0, \mathrm{b}))} \left\{ \frac{w^{(q)}(x; 0)}{w^{(q)}(a; 0)} w^{(q)}(a; y) - w^{(q)}(x; y) \right\} \mathrm{d}y. \end{aligned} \tag{8.5}$$

Although appealing to relatively straightforward methods, the proofs are quite long, requiring a little patience.

Proof of Theorem 8.2 Write $p(x,\delta) = \mathbb{E}_x(\mathrm{e}^{-q\kappa_a^+}\mathbf{1}_{(\kappa_a^+<\kappa_0^-)})$. Suppose that $x \le \mathrm{b}$. Then, by conditioning on $\mathcal{F}_{\tau_\mathrm{b}^+}$, we have

$$p(x,\delta) = \mathbb{E}_x\big(\mathrm{e}^{-q\tau_\mathrm{b}^+}\mathbf{1}_{(\tau_0^->\tau_\mathrm{b}^+)}\big)p(\mathrm{b},\delta) = \frac{W^{(q)}(x)}{W^{(q)}(\mathrm{b})}p(\mathrm{b},\delta), \tag{8.6}$$

where, in the last equality, we have used Theorem 4.5. Suppose now that $\mathrm{b} \le x \le a$. Using, Theorem 4.5, the strong Markov property (8.6) and the Gerber–Shiu measure from Theorem 5.5, we have

$$\begin{aligned} p(x,\delta) &= \mathrm{E}_x\big(\mathrm{e}^{-q\tau_a^+}\mathbf{1}_{(\tau_\mathrm{b}^->\tau_a^+)}\big) + \mathrm{E}_x\big(\mathrm{e}^{-q\tau_\mathrm{b}^-}\mathbf{1}_{(\tau_\mathrm{b}^-<\tau_a^+)}p(Z_{\tau_\mathrm{b}^-},\delta)\big) \\ &= \frac{\mathbb{W}^{(q)}(x-\mathrm{b})}{\mathbb{W}^{(q)}(a-\mathrm{b})} + \frac{p(\mathrm{b},\delta)}{W^{(q)}(\mathrm{b})}\mathrm{E}_x\big(\mathrm{e}^{-q\tau_\mathrm{b}^-}\mathbf{1}_{(\tau_\mathrm{b}^-<\tau_a^+)}W^{(q)}(X^\delta_{\tau_\mathrm{b}^-})\big) \\ &= \frac{\mathbb{W}^{(q)}(x-\mathrm{b})}{\mathbb{W}^{(q)}(a-\mathrm{b})} + \frac{p(\mathrm{b},\delta)}{W^{(q)}(\mathrm{b})}h(a,\mathrm{b},x), \end{aligned} \tag{8.7}$$

where

$$\begin{aligned} &h(a,\mathrm{b},x) \\ &= \int_0^{a-\mathrm{b}}\int_{(y,\infty)} W^{(q)}(\mathrm{b}+y-\theta) \\ &\qquad \times\left[\frac{\mathbb{W}^{(q)}(x-\mathrm{b})\mathbb{W}^{(q)}(a-\mathrm{b}-y)}{\mathbb{W}^{(q)}(a-\mathrm{b})} - \mathbb{W}^{(q)}(x-\mathrm{b}-y)\right]F(\mathrm{d}\theta)\mathrm{d}y. \end{aligned}$$

By setting $x = \mathrm{b}$ in (8.7) and recalling from Lemma 6.3 that $\mathbb{W}^{(q)}(0) = 1/(\mathrm{c}-\delta)$, we can now solve for $p(\mathrm{b},\delta)$. Indeed, we have

$$\begin{aligned} p(\mathrm{b},\delta) = W^{(q)}(\mathrm{b})\Big\{&(\mathrm{c}-\delta)\mathbb{W}^{(q)}(a-\mathrm{b})W^{(q)}(\mathrm{b}) \\ &- \int_0^{a-\mathrm{b}}\int_{(y,\infty)} W^{(q)}(\mathrm{b}+y-\theta) \\ &\qquad \times \mathbb{W}^{(q)}(a-\mathrm{b}-y)F(\mathrm{d}\theta)\mathrm{d}y\Big\}^{-1}. \end{aligned} \tag{8.8}$$

Next, we want to simplify the term involving the double integral in the above expression.

To this end, note that when $\delta = 0$ (the case that there is no refraction), we have, again by Theorem 4.5, that, for all $x \ge 0$,

$$p(\mathrm{b},0) = \mathbb{E}_\mathrm{b}\big(\mathrm{e}^{-q\tau_a^+}\mathbf{1}_{(\tau_0^->\tau_a^+)}\big) = \frac{W^{(q)}(\mathrm{b})}{W^{(q)}(a)}. \tag{8.9}$$

It follows, by comparing (8.8) (for $\delta = 0$) with (8.9), that

$$\int_0^{a-\mathrm{b}}\int_{(y,\infty)} W^{(q)}(\mathrm{b}+y-\theta)W^{(q)}(a-\mathrm{b}-y)F(\mathrm{d}\theta)\mathrm{d}y$$
$$= \mathrm{c}W^{(q)}(\mathrm{b})W^{(q)}(a-\mathrm{b}) - W^{(q)}(a). \tag{8.10}$$

As $a \geq \mathrm{b}$ is taken arbitrarily, we may take Laplace transforms in a on the interval (b, ∞) of both sides of the above expression. Denote by $\mathcal{L}_\mathrm{b}$ the operator satisfying

$$\mathcal{L}_\mathrm{b} f[\lambda] := \int_\mathrm{b}^\infty \mathrm{e}^{-\lambda x} f(x)\mathrm{d}x,$$

for non-negative functions f and let $\lambda > \Phi(q)$. For the left-hand side of (8.10), with the help of Fubini's Theorem, we get

$$\int_\mathrm{b}^\infty \mathrm{e}^{-\lambda x}\int_0^\infty\int_{(y,\infty)} W^{(q)}(\mathrm{b}+y+\theta)W^{(q)}(x-\mathrm{b}-y)\mathrm{d}yF(\mathrm{d}\theta)\mathrm{d}x$$
$$= \frac{\mathrm{e}^{-\lambda \mathrm{b}}}{\psi(\lambda)-q}\int_0^\infty\int_{(y,\infty)} \mathrm{e}^{-\lambda y}W^{(q)}(\mathrm{b}+y-\theta)F(\mathrm{d}\theta)\mathrm{d}y.$$

For the right-hand side of (8.10), we get

$$\int_\mathrm{b}^\infty \mathrm{e}^{-\lambda x}\big(W^{(q)}(x-\mathrm{b})\mathrm{c}W^{(q)}(\mathrm{b}) - W^{(q)}(x)\big)\mathrm{d}x$$
$$= \frac{\mathrm{e}^{-\lambda \mathrm{b}}}{\psi(\lambda)-q}\mathrm{c}W^{(q)}(\mathrm{b}) - \int_\mathrm{b}^\infty \mathrm{e}^{-\lambda x}W^{(q)}(x)\mathrm{d}x,$$

and so

$$\int_0^\infty\int_{(y,\infty)} \mathrm{e}^{-\lambda y}W^{(q)}(\mathrm{b}+y-\theta)F(\mathrm{d}\theta)\mathrm{d}y$$
$$= \mathrm{c}W^{(q)}(\mathrm{b}) - \big(\psi(\lambda)-q\big)\mathrm{e}^{\lambda \mathrm{b}}\mathcal{L}_\mathrm{b}W^{(q)}[\lambda], \tag{8.11}$$

for $\lambda > \Phi(q)$. Our objective now is to use (8.11) and show that, for $q \geq 0$ and $x \geq \mathrm{b}$, we have

$$\int_0^\infty\int_{(y,\infty)} W^{(q)}(\mathrm{b}+y-\theta)\mathbb{W}^{(q)}(x-\mathrm{b}-y)F(\mathrm{d}\theta)\mathrm{d}y$$
$$= -W^{(q)}(x) + (\mathrm{c}-\delta)W^{(q)}(\mathrm{b})\mathbb{W}^{(q)}(x-\mathrm{b})$$
$$-\delta\int_\mathrm{b}^x \mathbb{W}^{(q)}(x-y)W^{(q)\prime}(y)\mathrm{d}y. \tag{8.12}$$

We will do this by taking Laplace transforms in x on (b, ∞) on both sides of the equality in (8.12). To this end, note that, by (8.11), by Fubini's Theorem, the Laplace transform of the left-hand side of (8.12) is

$$\int_{\mathrm{b}}^{\infty} \mathrm{e}^{-\lambda x} \int_0^{\infty} \int_{(y,\infty)} W^{(q)}(\mathrm{b}+y-\theta)\mathbb{W}^{(q)}(x-\mathrm{b}-y)F(\mathrm{d}\theta)\mathrm{d}y\mathrm{d}x$$

$$= \frac{\mathrm{e}^{-\lambda \mathrm{b}}}{\psi(\lambda)-\delta\lambda-q}\big(\mathrm{c}W^{(q)}(\mathrm{b}) - \big(\psi(\lambda)-q\big)\mathrm{e}^{\lambda\mathrm{b}}\mathcal{L}_{\mathrm{b}}W^{(q)}[\lambda]\big), \tag{8.13}$$

where $\lambda > \varphi(q)$ and, for $q \geq 0$, $\varphi(q) = \sup\{\theta \geq 0 : \psi(\theta) - \mathrm{c}\theta = q\}$. (Note that φ is the right inverse of the Laplace exponent of X^{δ}.) Since

$$\mathcal{L}_{\mathrm{b}}\bigg(\int_{\mathrm{b}}^{x} f(x-y)g(y)\mathrm{d}y\bigg)[\lambda] = (\mathcal{L}_0 f)[\lambda](\mathcal{L}_{\mathrm{b}} g)[\lambda]$$

and, for $\lambda > \Phi(q)$,

$$\mathcal{L}_{\mathrm{b}}W^{(q)\prime}[\lambda] = \lambda\mathcal{L}_{\mathrm{b}}W^{(q)}[\lambda] - \mathrm{e}^{-\lambda\mathrm{b}}W^{(q)}(\mathrm{b})$$

(which follows from integration by parts), we have that the Laplace transform of the right-hand side of (8.12) is equal to the right-hand side of (8.13), for all sufficiently large λ. Hence (8.12) holds for almost every $x \geq \mathrm{b}$. Because both sides of (8.12) are continuous in x, we finally conclude that (8.12) holds for all $x \geq \mathrm{b}$.

To complete the proof, it suffices to plug (8.12) and the expression for $h(a, \mathrm{b}, x)$ into (8.7) and the desired identity follows after straightforward algebra. □

In anticipation of the proof of Theorem 8.3, we note here a particular identity which follows easily from (8.12). That is, for $v \geq u \geq m \geq 0$,

$$\int_0^{\infty} \int_{(z,\infty)} W^{(q)}(z-\theta+m)$$
$$\times \left[\frac{\mathbb{W}^{(q)}(v-m-z)}{\mathbb{W}^{(q)}(v-m)}\mathbb{W}^{(q)}(u-m) - \mathbb{W}^{(q)}(u-m-z)\right] F(\mathrm{d}\theta)\mathrm{d}z$$
$$= -\frac{\mathbb{W}^{(q)}(u-m)}{\mathbb{W}^{(q)}(v-m)}\bigg(W^{(q)}(v) + \delta\int_m^v \mathbb{W}^{(q)}(v-z)W^{(q)\prime}(z)\mathrm{d}z\bigg)$$
$$+ W^{(q)}(u) + \delta\int_m^u \mathbb{W}^{(q)}(u-z)W^{(q)\prime}(z)\mathrm{d}z. \tag{8.14}$$

Proof of Theorem 8.3 For Borel $B \subseteq [0, a]$ and $x, q \geq 0$, define

$$V^{(q)}(x, a, B) = \int_0^{\infty} \mathrm{e}^{-qt}\mathbb{P}_x\big(Z_t \in B, t < \kappa_0^- \wedge \kappa_a^+\big)\mathrm{d}t.$$

For $x \leq \mathrm{b}$, by the strong Markov property, Theorem 4.5 and Theorem 5.2, we have

$$V^{(q)}(x, a, B) = \mathbb{E}_x\bigg(\int_0^{\tau_{\mathrm{b}}^+} \mathrm{e}^{-qt}\mathbf{1}_{(Z_t\in B, t<\kappa_a^+\wedge\kappa_0^-)}\mathrm{d}t\bigg)$$
$$+ \mathbb{E}_x\bigg(\int_{\tau_{\mathrm{b}}^+}^{\infty} \mathrm{e}^{-qt}\mathbf{1}_{(Z_t\in B, t<\kappa_a^+\wedge\kappa_0^-, \tau_{\mathrm{b}}^+<\tau_0^-)}\mathrm{d}t\bigg)$$

$$= \mathbb{E}_x\left(\int_0^{\tau_b^+\wedge\tau_0^-} e^{-qt}\mathbf{1}_{(X_t\in B)}dt\right)$$

$$+\mathbb{E}_x\left(e^{-q\tau_b^+}\mathbf{1}_{(\tau_b^+<\tau_0^-)}\right)V^{(q)}(b,a,B)$$

$$= \int_B\left(\frac{W^{(q)}(b-y)}{W^{(q)}(b)}W^{(q)}(x) - W^{(q)}(x-y)\right)dy$$

$$+\frac{W^{(q)}(x)}{W^{(q)}(b)}V^{(q)}(b,a,B). \tag{8.15}$$

Moreover, for $b \le x \le a$, we have, using similar arguments,

$$V^{(q)}(x,a,B)$$

$$= \int_0^\infty e^{-qt}P_x\left(X_t^\delta \in B\cap[b,a], t<\tau_b^-\wedge\tau_a^+\right)dt$$

$$+E_x\left(\mathbf{1}_{(\tau_b^-<\tau_a^+)}e^{-q\tau_b^-}V^{(q)}\left(X^\delta_{\tau_b^-},a,B\right)\right)$$

$$= \int_{B\cap[b,a]}\left(\frac{\mathbb{W}^{(q)}(a-z)}{\mathbb{W}^{(q)}(a-b)}\mathbb{W}^{(q)}(x-b) - \mathbb{W}^{(q)}(x-z)\right)dz$$

$$+\int_0^\infty\int_{\theta<-z}\left\{\int_B\left[\frac{W^{(q)}(b-y)}{W^{(q)}(b)}W^{(q)}(z+\theta+b) - W^{(q)}(z+\theta+b-y)\right]dy\right.$$

$$\left.+\frac{V^{(q)}(b,a,B)}{W^{(q)}(b)}W^{(q)}(z+\theta+b)\right\}$$

$$\times\left[\frac{\mathbb{W}^{(q)}(a-b-z)}{\mathbb{W}^{(q)}(a-b)}\mathbb{W}^{(q)}(x-b) - \mathbb{W}^{(q)}(x-b-z)\right]F(d\theta)dz,$$

where in the first equality we have used the strong Markov property and in the second equality we have again used the Gerber–Shiu measure from Theorem 5.5. Next, we apply the identity (8.14) twice in order to simplify the expression for $V^{(q)}(x,a,B)$, $a \ge x \ge b$. We use it once by setting $m = b$, $u = x$, $v = a$ and once by setting $m = b - y$ and $u = x - y$, $v = a - y$ for $y \in [0, b]$. We obtain

$$V^{(q)}(x,a,B)$$

$$= \int_{B\cap[b,a]}\left(\frac{\mathbb{W}^{(q)}(a-z)}{\mathbb{W}^{(q)}(a-b)}\mathbb{W}^{(q)}(x-b) - \mathbb{W}^{(q)}(x-z)\right)dz$$

$$+\int_{B\cap[0,b)}\left\{\frac{W^{(q)}(b-y)}{W^{(q)}(b)}\left(-\frac{\mathbb{W}^{(q)}(x-b)}{\mathbb{W}^{(q)}(a-b)}w^{(q)}(a;0)+w^{(q)}(x;0)\right)\right.$$

$$\left.-\left(-\frac{\mathbb{W}^{(q)}(x-b)}{\mathbb{W}^{(q)}(a-b)}w^{(q)}(a;y)+w^{(q)}(x;y)\right)\right\}dy$$

$$+\frac{V^{(q)}(\mathrm{b},a,B)}{W^{(q)}(\mathrm{b})}\left(-\frac{\mathbb{W}^{(q)}(x-\mathrm{b})}{\mathbb{W}^{(q)}(a-\mathrm{b})}w^{(q)}(a;0)+w^{(q)}(x;0)\right). \tag{8.16}$$

Setting $x=\mathrm{b}$ in (8.16), we get an expression for $V^{(q)}(\mathrm{b},a,B)$ in terms of itself. Solving this and then putting the resulting expression for $V^{(q)}(\mathrm{b},a,B)$ back in (8.15) and (8.16) leads to (8.5), which completes the proof. □

8.3 Resolvent Density with Ruin

The two expressions we are interested in, namely the ruin probability and the expected present value of dividends paid until ruin, can both be extracted from the identity for the resolvent measure of Z when killed on exiting $[0,\infty)$,

$$\int_0^\infty \mathrm{e}^{-qt}\mathbb{P}_x\big(Z_t\in B, t<\kappa_0^-\big)\mathrm{d}t=\lim_{a\to\infty}V^{(q)}(x,a,B),$$

where B is any Borel set in $[0,\infty)$ and $q\geq 0$. Note that the limit is justified by monotone convergence. Recall that φ was defined as the right inverse of the Laplace exponent of X^δ, so that

$$\varphi(q)=\sup\big\{\theta\geq 0:\psi(\theta)-\delta\theta=q\big\}.$$

Corollary 8.4 *For $x,y,\mathrm{b}\geq 0$ and $q\geq 0$,*

$$\begin{aligned}
&\int_0^\infty \mathrm{e}^{-qt}\mathbb{P}_x\big(Z_t\in \mathrm{d}y, t<\kappa_0^-\big)\mathrm{d}t\\
&\quad=\mathbf{1}_{(y\in[\mathrm{b},\infty))}\left\{\frac{w^{(q)}(x;0)}{\delta\int_\mathrm{b}^\infty \mathrm{e}^{-\varphi(q)z}W^{(q)\prime}(z)\mathrm{d}z}\mathrm{e}^{-\varphi(q)y}-\mathbb{W}^{(q)}(x-y)\right\}\mathrm{d}y\\
&\qquad+\mathbf{1}_{(y\in[0,\mathrm{b}))}\left\{\frac{\int_\mathrm{b}^\infty \mathrm{e}^{-\varphi(q)z}W^{(q)\prime}(z-y)\mathrm{d}z}{\int_\mathrm{b}^\infty \mathrm{e}^{-\varphi(q)z}W^{(q)\prime}(z)\mathrm{d}z}w^{(q)}(x;0)-w^{(q)}(x;y)\right\}\mathrm{d}y. \qquad (8.17)
\end{aligned}$$

Proof Assume that $q>0$. We begin by noting that, from the representation of $\mathbb{W}^{(q)}$ in (4.4), it is straightforward to deduce that, for all $x,q>0$,

$$\lim_{a\to\infty}\frac{\mathbb{W}^{(q)}(a-x)}{\mathbb{W}^{(q)}(a)}=\mathrm{e}^{-\varphi(q)x}. \tag{8.18}$$

Note that, for each $q\geq 0$, $\varphi(q)\geq\Phi(q)$ and hence, appealing to the same representation in (4.4) for both $W^{(q)}$ and $\mathbb{W}^{(q)}$, it also follows that, for all $q,x>0$,

$$\lim_{a\to\infty}\frac{W^{(q)}(a-x)}{\mathbb{W}^{(q)}(a)}=0. \tag{8.19}$$

For $q > 0$, the result we are after is obtained by dividing the numerator and denominator of each of the first terms in the curly brackets of (8.5) by $\mathbb{W}^{(q)}(a)$ and taking limits as $a \to \infty$, making use of (8.18) and (8.19). The case that $q = 0$ is handled by taking limits as $q \downarrow 0$ in (8.17). □

Now we are in a position to derive expressions for (8.2) and (8.3).

Corollary 8.5 *For $x \geq 0$, if $\psi'(0+) \leq \delta$, then*

$$\mathbb{P}_x\left(\kappa_0^- < \infty\right) = 1.$$

Otherwise, when $\psi'(0+) > \delta$, we have

$$\mathbb{P}_x\left(\kappa_0^- < \infty\right) = 1 - \frac{\psi'(0+) - \delta}{1 - \delta W(\mathrm{b})}\left(W(x) + \delta \mathbf{1}_{(x \geq b)} \int_{\mathrm{b}}^{x} \mathbb{W}(x-y) W'(y) \mathrm{d}y\right). \tag{8.20}$$

Proof Let $\underline{Z}_t = \inf_{s \leq t} Z_s$ and, as usual, $\mathbf{e}_q$ denotes an independent and exponentially distributed random variable with mean $1/q$. Note that, for $q > 0$,

$$\begin{aligned}\mathbb{E}_x\left(\mathrm{e}^{-q\kappa_0^-}\mathbf{1}_{(\kappa_0^- < \infty)}\right) &= 1 - \mathbb{P}_x(\underline{Z}_{\mathbf{e}_q} \geq 0)\\ &= 1 - q\int_0^\infty \mathrm{e}^{-qt}\mathbb{P}_x\left(Z_t \in [0,\infty), t < \kappa_0^-\right)\mathrm{d}t.\end{aligned}$$

Computing the integral above from (8.17) is relatively straightforward and gives us, for $x, \mathrm{b} \geq 0$ and $q > 0$,

$$\begin{aligned}&\mathbb{E}_x\left(\mathrm{e}^{-q\kappa_0^-}\mathbf{1}_{(\kappa_0^- < \infty)}\right)\\ &= z^{(q)}(x) - \frac{q\int_{\mathrm{b}}^\infty \mathrm{e}^{-\varphi(q)y} W^{(q)}(y)\mathrm{d}y}{\int_{\mathrm{b}}^\infty \mathrm{e}^{-\varphi(q)y} W^{(q)\prime}(y)\mathrm{d}y} w^{(q)}(x;0)\\ &\quad + q\mathbf{1}_{(x\geq \mathrm{b})}\int_{\mathrm{b}}^x \mathbb{W}^{(q)}(x-z)\mathrm{d}z + q\int_0^{\mathrm{b}} W^{(q)}(x-z)\mathrm{d}z\\ &\quad - q\int_0^x W^{(q)}(z)\mathrm{d}z - q\delta\mathbf{1}_{(x\geq \mathrm{b})}\int_{\mathrm{b}}^x \mathbb{W}^{(q)}(x-z)W^{(q)}(z-\mathrm{b})\mathrm{d}z,\end{aligned} \tag{8.21}$$

where

$$z^{(q)}(x) = Z^{(q)}(x) + \delta q\int_{\mathrm{b}}^x \mathbb{W}^{(q)}(x-z)W^{(q)}(z)\mathrm{d}z, \quad x \in \mathbb{R}, q \geq 0.$$

The details of the computation are left to the reader.

Although it is not immediately obvious, it turns out that the last four terms in (8.21) combine to make zero. Indeed, in the case that $x < \mathrm{b}$, this observation is straightforward, noting that the indicators preceding the two integrals from b to x are identically zero and the second of these four terms may be replaced by $\int_0^x W^{(q)}(x-z)\mathrm{d}z = \int_0^x W^{(q)}(z)\mathrm{d}z$ on account of the fact that $W^{(q)}$ is identically zero on $(-\infty, 0)$. In the case that $x \geq \mathrm{b}$, the last four terms of (8.21) can be easily rearranged to give $C(x-\mathrm{b})$, where $C : [0,\infty) \to [0,\infty)$ is the continuous function

$$C(u) = q\int_0^u \mathbb{W}^{(q)}(z)\mathrm{d}z - q\int_0^u W^{(q)}(z)\mathrm{d}z - q\delta\int_0^u \mathbb{W}^{(q)}(z)W^{(q)}(u-z)\mathrm{d}z.$$

Taking Laplace transforms of C and using (4.1), we easily verify that C is identically zero.

In conclusion, we have that, for $x, \mathrm{b} \geq 0$ and $q > 0$,

$$\mathbb{E}_x\big(\mathrm{e}^{-q\kappa_0^-}\mathbf{1}_{(\kappa_0^-<\infty)}\big) = z^{(q)}(x) - \frac{q\int_{\mathrm{b}}^\infty \mathrm{e}^{-\varphi(q)y}W^{(q)}(y)\mathrm{d}y}{\int_{\mathrm{b}}^\infty \mathrm{e}^{-\varphi(q)y}W^{(q)\prime}(y)\mathrm{d}y}w^{(q)}(x;0).$$

The expression for the ruin probability in (8.20) can be obtained by taking limits on the left- and right-hand side above as $q \downarrow 0$. On the left-hand side, thanks to monotone convergence, the limit is equal to $\mathbb{P}_x(\kappa_0^- < \infty)$. Computing the limits on the right-hand side is relatively straightforward, since

$$\int_0^\infty \mathrm{e}^{-\varphi(q)z}\mathbb{W}^{(q)}(z)\mathrm{d}z = \frac{1}{\varphi(q)\delta} \tag{8.22}$$

and

$$\lim_{q\downarrow 0}\frac{q}{\varphi(q)} = \lim_{q\downarrow 0}\frac{\psi(\varphi(q)) - \delta\varphi(q)}{\varphi(q)} = 0 \vee \big(\psi'(0+) - \delta\big).$$

The details are again left to the reader. □

Corollary 8.6 *For $x, q, \mathrm{b} \geq 0$,*

$$\mathbb{E}_x\left(\int_0^{\kappa_0^-}\mathrm{e}^{-qt}\delta\mathbf{1}_{(Z_t>\mathrm{b})}\mathrm{d}t\right) = -\delta\mathbf{1}_{(x\geq\mathrm{b})}\int_{\mathrm{b}}^x \mathbb{W}^{(q)}(z-\mathrm{b})\mathrm{d}z$$
$$+\frac{W^{(q)}(x) + \delta\mathbf{1}_{(x\geq\mathrm{b})}\int_{\mathrm{b}}^x \mathbb{W}^{(q)}(x-y)W^{(q)\prime}(y)\mathrm{d}y}{\varphi(q)\int_0^\infty \mathrm{e}^{-\varphi(q)y}W^{(q)\prime}(y+b)\mathrm{d}y}.$$

Proof The proof is a simple exercise in integrating the resolvent measure (8.17) over (b,∞). □

8.4 Comments

The terminology "refraction" comes from Gerber and Shiu (2006b). See also Gerber and Shiu (2006a). As indicated in the introduction, refraction strategies emerge as an optimal solution to (1.7). Kyprianou et al. (2012) offers a general perspective on how to choose the refraction threshold b optimally for the setting that X is a spectrally negative Lévy process. Making use of Corollary 8.6, they introduce the function

$$r(\mathrm{b}) = \varphi(q) \int_0^x \mathrm{e}^{-\varphi(q)y} W^{(q)\prime}(y+\mathrm{b})\mathrm{d}y, \quad \mathrm{b} \geq 0.$$

Note in particular, from the same corollary, the expected present value of dividends until ruin with refraction threshold $\mathrm{b} \geq 0$ is equal to $W^{(q)}(x)/r(\mathrm{b})$, when $Z_0 = x \leq \mathrm{b}$. With a similar flavour to the case of reflection strategies, Kyprianou et al. (2012) give sufficient conditions under which the optimal threshold b should be chosen to minimise the function $r(\mathrm{b})$.

The majority of the arguments in this chapter are taken from Kyprianou and Loeffen (2010), where X is taken to be a general spectrally negative Lévy process. The question of existence and uniqueness for the SDE (8.1) when X is a general Lévy process has not yet been handled in the literature. In particular, when X has no Gaussian component, although still a simple-looking SDE, (8.1) presents some difficulties when levelled against standard theory. Further fluctuation identities for refracted spectrally negative Lévy processes can be found in Kyprianou et al. (2013).

Chapter 9
Concluding Discussion

On the one hand, the use of scale functions would appear to have made many of the problems we have looked at solvable. On the other hand, one may question the extent to which we have solved the posed problems as our scale functions are only defined in terms of a Laplace transform. We have arguably only provided a solution “up to the inversion of a Laplace transform”. It would be nice to have some concrete examples of scale functions. It turns out that few concrete examples are known and they are quite difficult to produce in general. Nonetheless, we shall show that there is still sufficient analytical structure known for a general scale function to justify their use, in particular when moving to the bigger class of processes for which the surplus process is modelled by a general spectrally negative Lévy process.

9.1 Mixed-Exponential Claims

In general, it is quite hard to construct the scale function $W^{(q)}$ for a Cramér–Lundberg process. There are a handful of claim distributions F, for which there is a reasonable degree of tractability as far as scale functions are concerned. One such example is the case that F belongs to the class of *mixed-exponential distributions*. This family is also known as the *hyper-exponential distributions*. In order to specify these distributions, let us define the arrival rate

$$\lambda = \sum_{j=1}^{m} a_j/\rho_j \tag{9.1}$$

and the claims distribution by

$$F(\mathrm{d}x) = \frac{1}{\lambda}\left(\sum_{j=1}^{m} a_j \mathrm{e}^{-\rho_j x}\right)\mathrm{d}x, \quad x > 0, \tag{9.2}$$

A.E. Kyprianou, *Gerber–Shiu Risk Theory*, EAA Series,
DOI 10.1007/978-3-319-02303-8_9,

where $m \in \mathbb{N}$, and, for $j = 1, \ldots, m$, the coefficients a_j and ρ_j are strictly positive. For convenience, we also assume that the ρ_j are arranged in increasing order. The premium rate c may be taken valued in $(0, \infty)$ without restriction.

It is easily verified that the Laplace exponent is given by

$$\psi(\theta) = \mathrm{c}\theta - \theta \sum_{j=1}^{m} \frac{a_j}{\rho_j(\rho_j + \theta)}, \qquad \theta \geq 0.$$

A little thought reveals that the exponent ψ is, in fact, well defined as a mapping from $\mathbb{C}$ into $\mathbb{C}$, provided one is prepared to see it as a meromorphic function which has only singular poles at $z = -\rho_j$. Henceforth, we shall treat ψ in the broader sense of a complex-valued function. We know from our previous analysis that, for $q > 0$, the equation $\psi(\theta) = q$ has a unique solution $z = \Phi(q)$ in the half-plane $\mathrm{Re}(\theta) > 0$, and we also know that this solution is real. It can be proved that if we look for other roots of the equation $\psi(\theta) = q$ on the complex plane, then there are precisely m of them and they are all to be found on the negative part of the real axis. If we write these solutions as $\theta = -\zeta_j$, for $j = 1, \ldots, m$, then it also turns out that they satisfy the interlacing property

$$0 < \zeta_1 < \rho_1 < \zeta_2 < \rho_2 < \ldots < \zeta_m < \rho_m. \tag{9.3}$$

The following lemma gives an explicit formula for the scale function expressed in terms of the roots ζ_j and the first derivative of the Laplace exponent.

Lemma 9.1 *If X is a Cramér–Lundberg process with λ and F given by* (9.1) *and* (9.2), *respectively, then, for all $q > 0$, the scale function is given by*

$$W^{(q)}(x) = \frac{\mathrm{e}^{\Phi(q)x}}{\psi'(\Phi(q))} + \sum_{j=1}^{m} \frac{\mathrm{e}^{-\zeta_j x}}{\psi'(-\zeta_j)}, \qquad x \geq 0. \tag{9.4}$$

Proof We sketch the proof. Knowing that $\{-\zeta_m, \ldots, -\zeta_1, \Phi(q)\}$ are all roots of the equation $\psi(\theta) = q$ in $\mathbb{C}$, the basic idea is to use partial fractions to write

$$\frac{1}{\psi(\theta) - q} = \frac{c_0}{(\theta - \Phi(q))} + \sum_{j=1}^{m} \frac{c_j}{(\theta + \zeta_j)}, \qquad \theta \in \mathbb{C}. \tag{9.5}$$

To determine the constant c_0, note that, for example,

$$\frac{1}{\psi'(\Phi(q))} = \lim_{\theta \to \Phi(q)} \frac{(\theta - \Phi(q))}{\psi(\theta) - q} = c_0 + \sum_{j=1}^{m} \lim_{\theta \to \Phi(q)} c_j \frac{(\theta - \Phi(q))}{(\theta + \zeta_j)} = c_0.$$

One derives $c_1, \ldots, c_m$ similarly. The identity (9.4) now follows by inverting (9.5) in a straightforward way. □

9.2 Spectrally Negative Lévy Processes

One of the advantages working with scale functions is that all of the results, as well as many of their proofs, are practically identical if we replace the Cramér–Lundberg process by a general spectrally negative Lévy process. Recall from Chap. 1 that X is a spectrally negative Lévy process if it has stationary independent increments, it has paths that are almost surely right-continuous with left limits, there are no positive discontinuities in its trajectories, and its paths are not monotone.

A simple example of a spectrally negative Lévy process is the resulting object we get from adding a Cramér–Lundberg process to a (scaled) independent Brownian motion, i.e.

$$X_t = \sigma B_t + ct - \sum_{i=1}^{N_t} \xi_i, \quad t \geq 0, \tag{9.6}$$

where $\{B_t : t \geq 0\}$ is a standard independent Brownian motion and $\sigma > 0$. This process is often referred to as a *perturbed* Cramér–Lundberg process.

In general, spectrally negative Lévy processes can be characterised through their Laplace exponent, also denoted by ψ, which satisfies

$$\psi(\theta) := \frac{1}{t} \log \mathbb{E}\big[e^{\theta X_t}\big],$$

and is well defined for $\theta \geq 0$. The Lévy–Khintchine formula, which is normally cited for the characteristic exponent of a Lévy process, also identifies the Laplace exponent in its general form as

$$\psi(\theta) = a\theta + \frac{1}{2}\sigma^2\theta^2 + \int_{(0,\infty)} \big(e^{-\theta x} - 1 + \theta x \mathbf{1}_{(x<1)}\big) \Pi(dx), \quad \theta \geq 0, \tag{9.7}$$

where $a \in \mathbb{R}$, $\sigma^2 \geq 0$ and the so-called *Lévy measure* Π is a (not-necessarily finite) measure concentrated on $(0, \infty)$ which satisfies the integrability condition

$$\int_{(0,\infty)} \big(1 \wedge x^2\big) \Pi(dx) < \infty. \tag{9.8}$$

Although ψ looks more complicated than for the case of a Cramér–Lundberg process, its shape is essentially the same. Indeed, it is not difficult to show, again by differentiation, that ψ is a strictly convex function which satisfies $\psi(0) = 0$ and $\psi(\infty) = \infty$. Moreover, just as with the Cramér–Lundberg process, we have $\mathbb{E}(X_1) = \psi'(0+) \in [-\infty, \infty)$. Accordingly, we may also work with the right inverse of ψ,

$$\Phi(q) := \sup\big\{\theta \geq 0 : \psi(\theta) = q\big\}, \quad q \geq 0.$$

Just as is the case with Cramér–Lundberg processes, the quantity $\Phi(0)$ is strictly positive if and only if $\psi'(0+) < 0$ and otherwise, when $\psi'(0+) \geq 0$, it is zero. In this respect, Fig. 2.1 could equally depict the Laplace exponent of a general spectrally negative Lévy process.

We should note that the class of spectrally negative Lévy processes contains the class of Cramér–Lundberg processes. Indeed, this can be seen by taking $\Pi(\mathrm{d}x) = \lambda F(\mathrm{d}x)$ on $(0,\infty)$ and $\sigma = 0$. In that case, (9.7) can be written

$$\psi(\theta) = \left(a + \lambda \int_{(0,1)} x F(\mathrm{d}x)\right)\theta - \lambda \int_{(0,\infty)} \left(1 - \mathrm{e}^{-\theta x}\right) F(\mathrm{d}x), \quad \theta \geq 0.$$

The exclusion of monotone paths from the definition of spectrally negative Lévy processes would force us to take

$$\mathrm{c} := a + \lambda \int_{(0,1)} x F(\mathrm{d}x) > 0. \tag{9.9}$$

Describing the paths of the Lévy process $X = \{X_t : t \geq 0\}$ associated to ψ is not as straightforward as in the case of a Cramér–Lundberg process. It is clear that the quadratic term $a\theta + \sigma^2\theta^2/2$ is the result of an independent linear Brownian component $\{at + \sigma B_t : t \geq 0\}$ in X. The integral term in ψ can be written

$$\begin{aligned}
&\int_{(0,\infty)} \left(\mathrm{e}^{-\theta x} - 1 + \theta x \mathbf{1}_{(x<1)}\right) \Pi(\mathrm{d}x) \\
&\quad = \sum_{n\geq 1} \left\{ \mathrm{c}_n \theta - \lambda_n \int_{(2^{-n}, 2^{-(n-1)}]} \left(1 - \mathrm{e}^{-\theta x}\right) \frac{\Pi(\mathrm{d}x)}{\lambda_n} \right\} \\
&\quad\quad - \lambda_0 \int_{(0,\infty)} \left(1 - \mathrm{e}^{-\theta x}\right) \frac{\Pi(\mathrm{d}x)}{\lambda_0}, \quad \theta \geq 0,
\end{aligned} \tag{9.10}$$

where $\lambda_0 = \Pi((1,\infty))$ and, for $n \geq 1$,

$$\mathrm{c}_n = \int_{(2^{-n}, 2^{-(n-1)}]} x \Pi(\mathrm{d}x) \quad \text{and} \quad \lambda_n = \Pi\left((2^{-n}, 2^{-(n-1)}]\right).$$

The term for λ_n is finite on account of (9.8). If $\lambda_n = 0$ for some $n \geq 0$, then we should understand the relevant term on the right-hand side of (9.10) as absent.

The decomposition (9.10), known as the *Lévy–Itô decomposition*, gives us the intuitive understanding that, for the given triplet (a, σ, Π), the associated spectrally negative Lévy process may be seen as the independent sum of a linear Brownian component, a series of Cramér–Lundberg processes and the negative of a compound Poisson process. The special choice of c_n, for $n \geq 1$, means that each of the Cramér–Lundberg processes have zero mean (in fact they are martingales). Moreover the n-th Cramér–Lundberg process experiences jumps whose magnitude falls strictly into the interval $(2^{-n}, 2^{-(n-1)}]$. Meanwhile, the compound Poisson process experiences jumps which are of magnitude strictly greater than 1.

The resulting path of the superimposition of these processes can be quite varied. Indeed, over any finite time horizon, there will be an almost surely infinite (albeit countable) number of jumps if and only if Π is an infinite measure. Moreover, X has paths of bounded variation (almost surely over each finite time horizon) if and only if $\int_{(0,1)} x\Pi(\mathrm{d}x) < \infty$ and $\sigma = 0$. Finally, X is a Cramér–Lundberg process if and only if $\sigma = 0$ and Π has finite total mass.

For spectrally negative Lévy processes, we may also define scale functions $W^{(q)}(x)$, $q \geq 0$, $x \in \mathbb{R}$, in exactly the same way as we did for Cramér–Lundberg processes. In particular, for $q \geq 0$, $W^{(q)}(x) = 0$ for $x < 0$ and otherwise, on $[0, \infty)$, it is the unique right-continuous increasing function whose Laplace transform satisfies

$$\int_0^\infty \mathrm{e}^{-\theta x} W^{(q)}(x)\mathrm{d}x = \frac{1}{\psi(\theta) - q}, \qquad \theta > \Phi(q). \tag{9.11}$$

The majority of the main results presented in the previous chapters and indeed many of their proofs, are still valid when the setting of a Cramér–Lundberg process is replaced by a general spectrally negative Lévy process. We are thus brought again to the question of the existence of concrete examples of scale functions.

Not surprisingly, it is also difficult to find closed-form examples of scale functions for a spectrally negative Lévy process that is not a Cramér–Lundberg process. Here are a couple of related examples.

A spectrally negative α-stable process, for $\alpha \in (1,2)$, has Lévy measure

$$\Pi(\mathrm{d}x) = \frac{k_\alpha}{x^{1+\alpha}}\mathrm{d}x, \qquad x > 0,$$

where k_α is a constant that can be chosen appropriately so that

$$\psi(\theta) = \theta^\alpha, \qquad \theta \geq 0.$$

Note that $\psi'(0+) = 0$ and hence the α-stable process oscillates.

Denote by

$$\mathcal{E}_{\alpha,\beta}(x) = \sum_{n\geq 0} \frac{x^n}{\Gamma(n\alpha + \beta)}, \qquad x \in \mathbb{R}, \tag{9.12}$$

the two-parameter Mittag–Leffler function. It is characterised by a Fourier–Laplace transform. Specifically, for $\lambda \in \mathbb{R}$, $\theta \in \mathbb{C}$ and $\mathrm{Re}(\theta) > |\lambda|^{1/\alpha}$, we have

$$\int_0^\infty \mathrm{e}^{-\theta x} x^{\beta-1} \mathcal{E}_{\alpha,\beta}\big(\lambda x^\alpha\big)\mathrm{d}x = \frac{\theta^{\alpha-\beta}}{\theta^\alpha - \lambda}. \tag{9.13}$$

We recognise immediately that

$$W^{(q)}(x) = x^{\alpha-1}\mathcal{E}_{\alpha,\alpha}\big(qx^\alpha\big), \qquad q, x \geq 0.$$

Suppose that we look at the α-stable process under the Esscher transform (2.7). As alluded to above, the main part of this result is still valid for general spectrally negative Lévy processes. In particular, we note that the class of spectrally negative Lévy processes is closed under the Esscher transformation. For each $\gamma \geq 0$ and

$\alpha \in (1,2)$, if $(X, \mathbb{P})$ is a spectrally negative α-stable process, then $(X, \mathbb{P}^{\gamma})$ is a spectrally negative Lévy process with Laplace exponent

$$\psi_{\gamma}(\theta) = (\theta + \gamma)^{\alpha} - \gamma^{\alpha}, \quad \theta \geq -\gamma.$$

This process is also known as a *tempered stable process*. Like the α-stable process, it has no Gaussian component. Similarly to the conclusion of Theorem 2.3 for Cramér–Lundberg processes, the effect of the Esscher transform is to exponentially tilt the Lévy measure so that the new Lévy measure satisfies

$$\Pi_{\gamma}(\mathrm{d}x) = \frac{k_{\alpha}\mathrm{e}^{-\gamma x}}{x^{1+\alpha}}\mathrm{d}x, \quad x > 0.$$

Note also that $\psi_{\gamma}'(0+) = \psi'(\gamma) > 0$ and hence the process drifts to $+\infty$.

Just as above, we may derive, by inspection of (9.13), the following identity for $W_{\gamma}^{(q)}$, the q-scale function of $(X, \mathbb{P}^{\gamma})$:

$$W_{\gamma}^{(q)}(x) = \mathrm{e}^{-\gamma x} x^{\alpha-1} \mathcal{E}_{\alpha,\alpha}\big((q + \gamma^{\alpha})x^{\alpha}\big), \quad x \geq 0.$$

9.3 Analytic Properties of Scale Functions

Despite the fact that, for most spectrally negative Lévy process, we are unable to invert the Laplace transform (9.11), we can nonetheless get an understanding of the shape of the general scale function. Here is a summary of some of the known facts, most of which can be derived from the Laplace transform (9.11).

Continuity at the Origin For all $q \geq 0$,

$$W^{(q)}(0+) = \begin{cases} 0 & \text{if } \sigma > 0 \text{ or } \int_{(0,1)} x\Pi(\mathrm{d}x) = \infty \\ \mathrm{c}^{-1} & \text{if } \sigma = 0 \text{ and } \int_{(0,1)} x\Pi(dx) < \infty, \end{cases} \tag{9.14}$$

where $\mathrm{c} = a + \int_{(-1,0)} x\Pi(dx)$.

Derivative at the Origin For all $q \geq 0$,

$$W_{+}^{(q)\prime}(0+) = \begin{cases} 2/\sigma^2 & \text{if } \sigma > 0 \\ \infty & \text{if } \sigma = 0 \text{ and } \Pi((0,\infty)) = \infty \\ [q + \Pi((0,\infty))]/\mathrm{c}^2 & \text{if } \sigma = 0 \text{ and } \Pi((0,\infty)) < \infty. \end{cases} \tag{9.15}$$

Behaviour at $+\infty$ for $q = 0$ As $x \to \infty$ we have

$$W(x) \sim \begin{cases} 1/\psi'(0+) & \text{if } \psi'(0+) > 0 \\ \mathrm{e}^{\Phi(0)x}/\psi'(\Phi(0)) & \text{if } \psi'(0+) < 0. \end{cases} \tag{9.16}$$

When $\mathbb{E}(X_1) = 0$ a number of different asymptotic behaviours may occur. For example, if $\phi(\theta) := \psi(\theta)/\theta$ satisfies $\phi'(0+) < \infty$ then $W(x) \sim x/\phi'(0+)$ as $x \to \infty$.

Behaviour at $+\infty$ for $q > 0$ As $x \to \infty$ we have

$$W^{(q)}(x) \sim e^{\Phi(q)x} / \psi'(\Phi(q)) \tag{9.17}$$

and thus there is asymptotic exponential growth.

Smoothness It is known that if X has paths of bounded variation, then, for all $q \geq 0$, $W^{(q)}|_{(0,\infty)} \in C^1(0,\infty)$ if and only if Π has no atoms. In the case that X has paths of unbounded variation, it is known that, for all $q \geq 0$, $W^{(q)}|_{(0,\infty)} \in C^1(0,\infty)$. Moreover if $\sigma > 0$, then $C^1(0,\infty)$ may be replaced by $C^2(0,\infty)$. Clearly this picture is incomplete.

Taking account of the fact that the Laplace transform of $W^{(q)}$ is expressed in terms of $\psi(\theta)$, which, itself, can be considered as a type of analytical transform of the measure Π, it is not surprising that there is an intimate connection between the smoothness of the scale functions and the Lévy measure. Whilst there are a number of existing results connecting the two (see Sect. 9.5), a general result remains at large.

Concavity and Convexity If $x \mapsto \overline{\Pi}(x)$, $x > 0$, is a completely monotone function,[1] then, for all $q > 0$, $W^{(q)\prime}(x)$, $x > 0$, is convex. Note in particular, the latter implies that there exists an $a^* \geq 0$ such that $W^{(q)}$ is concave on $(0, a^*)$ and convex on (a^*, ∞). In the case that $\psi'(0+) \geq 0$ and $q = 0$, the same conclusion holds with $a^* = \infty$, which is to say that W is a concave function. More generally, we have the following result.

Theorem 9.2 *Suppose that $\overline{\Pi}$ is log-convex. Then for all $q \geq 0$, $W^{(q)}$ has a log-convex first derivative.*

(Note that a completely monotone function is log-convex and a log-convex function is also convex.)

9.4 Engineered Scale Functions

Within the class of spectrally negative Lévy processes, there are a number of methods for generating examples of scale functions with $q = 0$. Rather than trying to invert (9.11) for a given ψ, the idea is to construct a ψ corresponding to a given W. We outline one method here, which is based around the *Wiener–Hopf factorisation.*

---

[1] A smooth function $f : (0,\infty) \to [0,\infty)$ is completely monotone if, for all $n \in \mathbb{N}$,

$$(-1)^n \frac{d^n f(x)}{dx^n} \geq 0.$$

For the purpose of this discussion, the Wiener–Hopf factorisation concerns the Laplace exponent of X and takes the form:

$$\psi(\theta) = \big(\theta - \Phi(0)\big)\phi(\theta), \quad \theta \geq 0, \tag{9.18}$$

where ϕ is a so-called *Bernstein function*. Specifically, the term $\phi(\theta)$ must necessarily take the form

$$\phi(\theta) = \kappa + \delta\theta + \int_{(0,\infty)} \big(1 - \mathrm{e}^{-\theta x}\big)\Upsilon(\mathrm{d}x), \quad \theta \geq 0, \tag{9.19}$$

where $\kappa, \delta \geq 0$ and Υ is a measure concentrated on $(0, \infty)$ which satisfies $\int_{(0,\infty)}(1 \wedge x)\Upsilon(\mathrm{d}x) < \infty$. Roughly speaking, this factorisation can be proved by recalling that $\psi(\Phi(0)) = 0$ and then manually factoring out $(\theta - \Phi(0))$ from ψ by using integration by parts to deal with the integral part of (9.7). It turns out that

$$\Upsilon\big((x,\infty)\big) = \mathrm{e}^{\Phi(0)x} \int_x^\infty \mathrm{e}^{-\Phi(0)u}\overline{\Pi}(u)\mathrm{d}u \quad \text{for } x > 0, \tag{9.20}$$

$\delta = \sigma^2/2$ and $\kappa = \psi'(0+) \vee 0$.

It is remarkable that ϕ is also the Laplace exponent of a subordinator[2] which is sent to the cemetery state $+\infty$ after an independent and exponentially distributed random time with rate κ.[3] If we write this process $H = \{H_t : t \geq 0\}$ and let $\zeta = \inf\{t > 0 : H_t = +\infty\}$, then, for all $t \geq 0$,

$$\phi(\theta) = -\frac{1}{t}\log E\big(\mathrm{e}^{-\theta H_t}\mathbf{1}_{(t<\zeta)}\big), \quad \theta \geq 0.$$

What is even more remarkable is that the range of $\{H_t : t < \zeta\}$ agrees precisely with the range of the process $\{-\underline{X}_t : t \geq 0\}$. Accordingly we call H the descending ladder height process of X.

In the special case that $\Phi(0) = 0$, that is to say, the process X does not drift to $-\infty$, or equivalently that $\psi'(0+) \geq 0$, it can be shown that the scale function W describes the renewal measure of H. Indeed, the renewal measure of H is defined by

$$\mathcal{U}(\mathrm{d}x) = \int_0^\infty \mathrm{d}t \cdot P(H_t \in \mathrm{d}x, t < \zeta), \quad \text{for } x \geq 0. \tag{9.21}$$

Calculating its Laplace transform we get the identity

$$\int_0^\infty \mathrm{e}^{-\theta x}\mathcal{U}(\mathrm{d}x) = \frac{1}{\phi(\theta)} \quad \text{for } \theta > 0. \tag{9.22}$$

---

[2] A subordinator is a Lévy process with non-decreasing paths.

[3] Recall our convention that an exponential random variable with rate 0 is defined to be infinite-valued with probability 1.

Recall that we can integrate (9.11) by parts and get

$$\int_{[0,\infty)} e^{-\theta x} W(dx) = \frac{\theta}{\psi(\theta)} = \frac{1}{\phi(\theta)}, \quad \theta \geq 0.$$

Hence, it appears that W agrees precisely with the renewal function, $\mathcal{U}$, of the subordinator H that appears in the Wiener–Hopf factorisation.

It can be shown similarly that, when $\Phi(0) > 0$, the scale function is related to the renewal measure of H by the formula

$$W(x) = e^{\Phi(0)x} \int_0^x e^{-\Phi(0)y} \mathcal{U}(dy), \quad x \geq 0. \tag{9.23}$$

This relationship between scale functions and renewal measures of subordinators lies at the heart of the approach we shall describe in this section for engineering scale functions. A key to the method is the fact that one can find in the literature several subordinators for which the renewal measure is known explicitly. Should these subordinators turn out to be the descending ladder height process of a spectrally negative Lévy process, then this would give an exact expression for its scale function. Said another way, we can build scale functions using the following approach.

Step 1. Choose a subordinator, say H, with Laplace exponent ϕ, for which one knows its renewal measure, $\mathcal{U}$, or equivalently, in light of (9.22), one can explicitly invert the Laplace transform $1/\phi(\theta)$.

Step 2. Choose a constant $\varphi \geq 0$ and verify whether the relation

$$\psi(\theta) := (\theta - \varphi)\phi(\theta), \quad \theta \geq 0,$$

defines the Laplace exponent of a spectrally negative Lévy process.

Step 3. Once Steps 1 and 2 are verified, then the scale function of the spectrally negative Lévy process we have generated is given by (9.23).

Of course, for this method to be useful we should first provide necessary and sufficient conditions for the pair (H, φ) to belong to the Wiener–Hopf factorisation of a spectrally negative Lévy process.

The following theorem shows how one may identify a spectrally negative Lévy process X (called the *parent process*) for a given pair (H, φ). The proof follows by a straightforward manipulation of the Wiener–Hopf factorisation (9.18).

Theorem 9.3 *Suppose that H is a subordinator, killed at rate $\kappa \geq 0$, with drift $\delta \geq 0$ and Lévy measure Υ which is absolutely continuous with non-increasing density. Suppose further that $\varphi \geq 0$ is given such that $\varphi\kappa = 0$. Then there exists a spectrally negative Lévy process X, henceforth referred to as the "parent process", whose descending ladder height process is precisely the process H. The Lévy triplet (a, σ, Π) of the parent process is uniquely identified as follows. The Gaussian coefficient is given by $\sigma = \sqrt{2\delta}$. The Lévy measure is given by*

$$\overline{\Pi}(x) = \varphi \Upsilon(x, \infty) + \frac{d\Upsilon}{dx}(x).$$

Finally, a can be chosen such that

$$\psi(\theta) = (\theta - \varphi)\phi(\theta), \tag{9.24}$$

for $\theta \geq 0$ *where* $\phi(\theta) = -\log \mathbb{E}(\mathrm{e}^{-\theta H_1})$.

Conversely, the killing rate, drift and Lévy measure of the descending ladder height process associated to a given spectrally negative Lévy process X are also given by the above formulae when one replaces φ *by* $\Phi(0)$.

Let us conclude this section by presenting a concrete example of how this methodology works in practise. Consider a spectrally negative Lévy process which is the parent process of a (killed) tempered stable subordinator. That is to say a subordinator with Laplace exponent given by

$$\phi(\theta) = \kappa - c\Gamma(-\alpha)\big((\gamma+\theta)^{\alpha} - \gamma^{\alpha}\big),$$

where $\alpha \in (-1, 1) \setminus \{0\}$, $\gamma > 0$, $\kappa \geq 0$ and $c > 0$. The Lévy measure corresponding to this subordinator satisfies

$$\Upsilon(\mathrm{d}x) = c\frac{\mathrm{e}^{-\gamma x}}{x^{\alpha+1}}\mathrm{d}x, \quad x > 0.$$

The corresponding Lévy measure of the parent process is given by

$$\Pi(\mathrm{d}x) = c\frac{(\varphi+\gamma)}{x^{\alpha+1}}\mathrm{e}^{-\gamma x}\mathrm{d}x + c\frac{(\alpha+1)}{x^{\alpha+2}}\mathrm{e}^{-\gamma x}\mathrm{d}x, \quad x > 0. \tag{9.25}$$

Note from Theorem 9.3 that $\sigma = 0$, indicating the absence of a Gaussian component.

If $0 < \alpha < 1$, then the jump component is the sum of the negative of an infinite activity tempered stable subordinator and an independent spectrally negative tempered stable process with infinite variation. If $-1 \leq \alpha < 0$, then the jump part of the parent process is the independent sum of the negative of a tempered stable subordinator with stability parameter $1 + \alpha$ and exponential parameter γ, and the negative of an independent compound Poisson subordinator with jumps from a gamma distribution shape parameter $-\alpha$ and rate parameter γ.

One easily deduces the following transformations as special examples of (9.13) for $\theta, \lambda > 0$, such that $|\theta^{\alpha}/\lambda| > 1$

$$\int_0^{\infty} \mathrm{e}^{-\theta x} x^{\alpha-1} \mathcal{E}_{\alpha,\alpha}\big(\lambda x^{\alpha}\big)\mathrm{d}x = \frac{1}{\theta^{\alpha} - \lambda}, \tag{9.26}$$

and

$$\int_0^{\infty} \mathrm{e}^{-\theta x} \lambda^{-1} x^{-\alpha-1} \mathcal{E}_{-\alpha,-\alpha}\big(\lambda^{-1} x^{-\alpha}\big)\mathrm{d}x = \frac{\lambda}{\lambda - \theta^{\alpha}} - 1, \tag{9.27}$$

valid for $\alpha > 0$ and $\alpha < 0$, respectively. Together with the well-known rules for Laplace transforms concerning primitives and tilting, we may quickly deduce the following expressions for the scale functions associated to the parent process with Laplace exponent given by (9.24) such that $\kappa\varphi = 0$.

If $0 < \alpha < 1$, then, for $x \geq 0$,

$$W(x) = \frac{e^{\varphi x}}{-c\Gamma(-\alpha)} \int_0^x e^{-(\gamma+\varphi)y} y^{\alpha-1} \mathcal{E}_{\alpha,\alpha}\left(\frac{\kappa + c\Gamma(-\alpha)\gamma^\alpha}{-c\Gamma(-\alpha)} y^\alpha\right) dy.$$

If $-1 < \alpha < 0$, then, for $x \geq 0$,

$$W(x) = \frac{e^{\varphi x}}{\kappa + c\Gamma(-\alpha)\gamma^\alpha} + \frac{c\Gamma(-\alpha)e^{\varphi x}}{(\kappa + c\Gamma(-\alpha)\gamma^\alpha)^2} \int_0^x e^{-(\gamma+\varphi)y} y^{-\alpha-1} \mathcal{E}_{-\alpha,-\alpha}\left(\frac{c\Gamma(-\alpha)y^{-\alpha}}{\kappa + c\Gamma(-\alpha)\gamma^\alpha}\right) dy.$$

9.5 Comments

The case of a Cramér–Lundberg process with mixed exponential jumps can be generalised by taking jumps whose distribution has a Laplace transform that is the ratio of two polynomial functions of finite degree (also called a rational Laplace transform). Another favourite class in the family of jump distributions with rational Laplace transform (which also contains the class of mixed-exponential distributions) is the one of phase-type distributions. Neuts (1981) gives an overview of the latter. The tractability of the class of processes with jumps having rational transform can be routed back to early work of Borovkov (1976) concerning the Wiener–Hopf factorisation. Many authors have worked on these types of Cramér–Lundberg processes and it would be impossible to give a complete list here. We cite instead three of the most recent references which give a good overview in the context of Gerber–Shiu type problems. These are Asmussen and Albrecher (2010), Kuznetsov and Morales (2011) and Egami and Yamazaki (2012). The idea of engineering scale functions through the Wiener–Hopf factorisation comes from Hubalek and Kyprianou (2010) and Kypianou and Rivero (2008). Analytical properties of scale functions have been described in a variety of papers. A recent summary of these and many more facts can be found in the review on scale functions found in Cohen et al. (2013).

References

Albrecher, H., Borst, S., Boxma, O., Resing, J.: Ruin excursions, the $G/G/\infty$ queue, and tax payments in renewal risk models. J. Appl. Probab. **48A**, 3–14 (2011)

Albrecher, H., Gerber, H.: A note on moments of dividends. Acta Mathematica Applicatae Sinica **27**, 353–354 (2011)

Albrecher, H., Hipp, C.: Lundberg's risk process with tax. Blaetter der DGVFM **28**, 13–28 (2007)

Albrecher, H., Renaud, J.-F., Zhou, X.: A Lévy insurance risk process with tax. J. Appl. Probab. **45**, 363–375 (2008)

Asmussen, S., Albrecher, H.: Ruin Probabilities, 2nd edn. Advanced Series on Statistical Science & Applied Probability, vol. 14. World Scientific, Hackensack (2010)

Asmussen, S., Taksar, M.: Controlled diffusion models for optimal dividend pay-out. Insurance Math. Econom. **20**, 1–15 (1997)

Avram, F., Palmowski, Z., Pistorius, M.R.: On the optimal dividend problem for a spectrally negative Lévy process. Ann. Appl. Probab. **17**, 156–180 (2007)

Azcue, P., Muler, N.: Optimal reinsurance and dividend distribution policies in the Cramér–Lundberg model. Math. Finance **15**, 261–308 (2005)

Azéma, J., Yor, M.: Une solution simple au problème de Skorokhod. In: Séminaire de Probabilités XIII. Lecture Notes in Mathematics, vol. 721, pp. 91–115. Springer, Berlin (1979)

Bertoin, J.: An extension of Pitman's theorem for spectrally positive Lévy processes. Ann. Probab. **20**, 1464–1483 (1992)

Bertoin, J.: Lévy Processes. Cambridge University Press, Cambridge (1996)

Bertoin, J.: Exponential decay and ergodicity of completely asymmetric Lévy processes in a finite interval. Ann. Appl. Probab. **7**, 156–169 (1997)

Bertoin, J., Yor, M.: Exponential functionals of Lévy processes. Probab. Surv. **2**, 191–212 (2005)

Biffis, E., Morales, M.: On a generalization of the Gerber–Shiu function to path-dependent penalties. Insurance Math. Econom. **46**, 92–97 (2010)

Bingham, N.H.: Fluctuation theory in continuous time. Adv. Appl. Probab. **7**, 705–766 (1975)

Borovkov, A.A.: Stochastic Processes in Queueing Theory. Springer, Berlin (1976)

Cohen, S., Kuznetsov, A., Kyprianou, A.E., Rivero, V.: Lévy Matters II. Lecture Notes in Mathematics, vol. 2061. Springer, Heidelberg (2013)

Cramér, H.: Collected Works, vol. I. Edited and with a preface by Anders Martin-Löf. Springer, Berlin (1994a)

Cramér, H.: Collected Works, vol. II. Edited and with a preface by Anders Martin-Löf. Springer, Berlin (1994b)

de Finetti, B.: Su un'impostazione alternativa della teoria collettiva del rischio. Transactions of the XVth International Congress of Actuaries **2**, 433–443 (1957)

Dickson, D.C.M.: Insurance Risk and Ruin. Cambridge University Press, Cambridge (1999)

Dickson, D.C.M., Waters, H.R.: Some optimal dividends problems. Astin Bull. **34**, 49–74 (2004)

A.E. Kyprianou, *Gerber–Shiu Risk Theory*, EAA Series,
DOI 10.1007/978-3-319-02303-8,

Egami, M., Yamazaki, K.: Phase-type fitting of scale functions for spectrally negative Lévy processes. Preprint (2012)
Embrechts, P., Klüppelberg, C., Mikosch, T.: Modelling Extremal Events for Insurance and Finance. Springer, Berlin (1997)
Esscher, F.: On the probability function in the collective theory of risk. Skandinavisk Aktuarietidskrift **15**, 175–195 (1932)
Feller, W.: An Introduction to Probability Theory and Its Applications, vol II, 2nd edn. Wiley, New York (1971)
Gerber, H.U.: Entscheidungskriterien für den zusammengesetzten Poisson-Prozess. Schweiz. Verein. Versicherungsmath. Mitt. **69**, 185–228 (1969)
Gerber, H.U.: Games of economic survival with discrete and continuous income processes. Operations Research **20**, 37–45 (1972)
Gerber, H.U.: When does the surplus reach a given target? Insurance Math. Econom. **9**, 115–119 (1990)
Gerber, H.U., Shiu, E.S.W.: Option pricing by Esscher transforms. Trans. - Soc. Actuar. **46**, 99–191 (1994)
Gerber, H.U., Shiu, E.S.W.: The joint distribution of the time of ruin, the surplus immediately before ruin, and the deficit at ruin. Insurance Math. Econom. **21**, 129–137 (1997)
Gerber, H.U., Shiu, E.S.W.: On the time value of ruin. North Am. Actuar. J. **2**, 48–78 (1998)
Gerber, H.U., Shiu, E.S.W.: On optimal dividends: From reflection to refraction. J. Comput. Appl. Math. **186**, 4–22 (2006a)
Gerber, H.U., Shiu, E.S.W.: On optimal dividend strategies in the compound Poisson model. North Am. Actuar. J. **10**, 76–93 (2006b)
Greenwood, P.E., Pitman, J.W.: Fluctuation identities for Lévy processes and splitting at the maximum. Adv. Appl. Probab. **12**, 839–902 (1980)
Hubalek, F., Kyprianou, A.E.: Old and new examples of scale functions for spectrally negative Lévy processes. In: Dalang, R., Dozzi, M., Russo, F. (eds.) Sixth Seminar on Stochastic Analysis, Random Fields and Applications. Progress in Probability, pp. 119–146. Birkhäuser, Basel (2010)
Itô, K.: Poisson point processes attached to Markov processes. In: Proceedings of the Sixth Berkeley Symposium on Mathematical Statistics and Probability (Univ. California, Berkeley, Calif., 1970/1971), Vol. III: Probability Theory, pp. 225–239. University of California Press, Berkeley (1972)
Jeanblanc, M., Shiryaev, A.N.: Optimization of the flow of dividends. Russian Math. Surveys **50**, 257–277 (1995)
Kella, O., Whitt, W.: Useful martingales for stochastic storage processes with Lévy input. J. Appl. Probab. **29**, 396–403 (1992)
Kallenberg, O.: Foundations of Modern Probability, 2nd edn. Probability and Its Applications. Springer, New York (2002)
Kennedy, D.: Some martingales related to cumulative sum tests and single-server queues. Stoch. Proc. Appl. **4**, 261–269 (1976)
Kingman, J.F.C.: Poisson Processes. Oxford University Press, Oxford (1993)
Korolyuk, V.S.: Boundary problems for a compound Poisson process. Theory Probab. Appl. **19**, 1–14 (1974)
Korolyuk, V.S.: Boundary Problems for Compound Poisson Processes. Naukova Dumka, Kiev (1975a)
Korolyuk, V.S.: On ruin problem for compound Poisson process. Theory Probab. Appl. **20**, 374–376 (1975b)
Korolyuk, V.S., Borovskich, Ju.V.: Analytic Problems of the Asymptotic Behaviour of Probability Distributions. Naukova Dumka, Kiev (1981)
Korolyuk, V.S., Suprun, V.N., Shurenkov, V.M.: Method of potential in boundary problems for processes with independent increments and jumps of the same sign. Theory Probab. Appl. **21**, 243–249 (1976)

Kuznetsov, A., Morales, M.: Computing the finite-time expected discounted penalty function for a family of Lévy risk processes. Scandinavian Actuarial Journal (2011). doi:10.1080/03461238.2011.627747

Kyprianou, A.E.: Fluctuations of Lévy Processes with Applications, Introductory Lectures, 2nd edn. Springer, Heidelberg (2013)

Kyprianou, A.E., Loeffen, R.L.: Refracted Lévy processes. Ann. Inst. H. Poincaré. **46**, 24–44 (2010)

Kyprianou, A.E., Loeffen, R.L., Pérez, J.L.: Optimal control with absolutely continuous strategies for spectrally negative Lévy processes. J. Appl. Probab. **49**, 150–166 (2012)

Kyprianou, A.E., Ott, C.: Spectrally negative Lévy processes perturbed by functionals of their running supremum. J. Appl. Probab. **49**, 1005–1014 (2012)

Kyprianou, A.E., Palmowski, Z.: Distributional study of de Finetti's dividend problem for a general Lévy insurance risk process. J. Appl. Probab. **44**, 428–443 (2007)

Kyprianou, A.E., Pardo, J.C., Pérez, J.L.: Occupation times of refracted Lévy processes. Journal of Theoretical Probability (2013). doi:10.1007/s10959-013-0501-4

Kypianou, A.E., Rivero, V.: Special, conjugate and complete scale functions for spectrally negative Lévy processes. Electron. J. Probab. **13**, 1672–1701 (2008)

Kyprianou, A.E., Zhou, X.: General tax structures and the Lévy insurance risk model. J. Appl. Probab. **46**, 1146–1156 (2009)

Kyprianou, A.E., Rivero, V., Song, R.: Convexity and smoothness of scale functions and de Finetti's control problem. J. Theoret. Probab. **23**, 547–564 (2010)

Loeffen, R.L.: On optimality of the barrier strategy in de Finetti's dividend problem for spectrally negative Lévy processes. Ann. Appl. Probab. **18**, 1669–1680 (2008)

Loeffen, R.L., Renaud, J.-F.: De Finetti's optimal dividends problem with an affine penalty function at ruin. Insurance Math. Econom. **46**, 98–108 (2010)

Lundberg, F.: Approximerad Framställning Av Sannolikhetsfunktionen. Återförsäkring Av Kollektivrisker. Akad. Afhandling. Almqvist. och Wiksell, Uppsala (1903)

Neuts, M.F.: Matrix-Geometric Solutions in Stochastic Models. An Algorithmic Approach. Johns Hopkins Series in the Mathematical Sciences, vol. 2. Johns Hopkins University Press, Baltimore (1981)

Renaud, J.-F., Zhou, X.: Moments of the expected present value of total dividends until ruin in a Lévy risk model. J. Appl. Probab. **44**, 420–427 (2007)

Revuz, D., Yor, M.: Continuous Martingales and Brownian Motion, 3rd edn. Springer, Berlin (2004)

Rogers, L.C.G.: A guided tour through excursions. Bull. London Math. Soc. **21**, 305–341 (1989)

Rogers, L.C.G.: The two-sided exit problem for spectrally positive Lévy processes. Adv. Appl. Probab. **22**, 486–487 (1990)

Schmidli, H.: Stochastic Control in Insurance. Probability and Its Applications. Springer, London (2008)

Suprun, V.N.: Problem of ruin and resolvent of terminating processes with independent increments. Ukrainian Math. J. **28**, 39–45 (1976)

Takács, L.: Combinatorial Methods in the Theory of Stochastic Processes. Wiley, New York (1966)

Wald, A.: On cumulative sums of random variables. Ann. Math. Stat. **15**, 283–296 (1944)

Wald, A.: Some generalizations of the theory of cumulative sums of random variables. Ann. Math. Stat. **16**, 287–293 (1945)

Zolotarev, V.M.: The first passage time of a level and the behaviour at infinity for a class of processes with independent increments. Theory Probab. Appl. **9**, 653–661 (1964)

MIX
Papier aus verantwortungsvollen Quellen
Paper from responsible sources
FSC® C105338

If you have any concerns about our products,
you can contact us on
ProductSafety@springernature.com

In case Publisher is established outside the EU,
the EU authorized representative is:
Springer Nature Customer Service Center GmbH
Europaplatz 3, 69115 Heidelberg, Germany

Printed by Libri Plureos GmbH
in Hamburg, Germany